Quantum Physics
for beginners

A simple guide for discovering the hidden side of reality.
Master the theory of relativity and the mechanics of particles like Einstein | With easy and practical examples!

Table of Contents

Chapter 1 A pleasant surprise 1

Chapter 2 The beginning 33

Chapter 3 What is light? 43

Chapter 4 Mr. Planck... 63

Chapter 5 An uncertain Heisenberg 83

Chapter 6 Quantum..107

Chapter 7 Einstein and Bohr 121

Chapter 8 Quantum Physics in Present Times.......................139

Chapter 9 The third millennium...................................167

A pleasant surprise

B efore the quantum era, science lived on decisive pronouncements on causes and effects of motions: well defined objects moved along precise trajectories, in response to the action of various forces. But the science that we now call classical, emerged from the mists of a long history and duration until the end of the nineteenth century, overlooked the fact that each object was actually made up of a gigantic number of atoms. In a grain of sand, for example, there are several billion of them.

Before the quantum era, who observed a phenomenon was like an alien from space, who looked at the Earth from above and noticed only the movements of large crowds of thousands and thousands of people. Maybe they saw them marching in compact ranks, or applauding, or hurrying to work, or scattering in the streets. But nothing they observed could ever prepare them for what they would see by focusing their attention on individuals. On an individual level, humans showed behavior that could not be deduced from that of crowds - things like laughter, affection, compassion and creativity. Aliens, perhaps robotic probes or evolved insects, may not have had the right words to describe what they saw when they observed us closely. On the other hand, even we, today, with all the literature and poetry accumulated over the millennia, sometimes we can not fully understand the individual experiences of other human beings.

At the beginning of the 20th century something similar happened. The complex building of physics, with its exact predictions about the behavior of objects, i.e. crowds of atoms, suddenly collapsed. Thanks to new, sophisticated experiments, conducted with great skill, it was possible to study the properties not only of individual atoms, but also of the smaller particles of which they were made. It was like going from listening to an orchestral ensemble to quartets, trios and solo pieces. And the atoms seemed to behave in a disconcerting way in the eyes of the greatest physicists of the time, who were awakening from the sleep of the classical age. They were explorers of an unprecedented world, the equivalent of the poetic, artistic and musical avant-garde of the time. Among them were the most famous: Heinrich Hertz, Ernest Rutherford, J. J. Thomson, Niels Bohr, Marie Curie, Werner Heisenberg, Erwin Schrödinger, Paul Dirac, Louis-Victor de Broglie, Albert Einstein, Max Born, Max Planck and Wolfgang Pauli. The shock they felt after poking around inside the atoms was equal to what the crew of the Enterprise must have experienced when they first encountered an alien civilization found in the vastness of the cosmos. The confusion produced by the examination of the new data slowly stimulated the first, desperate attempts by physicists to restore some order and logic in their science. At the end of the twenties of the last century the fundamental structure of the atom could be said by now known in broad lines, and it could be applied to the chemistry and physics of ordinary matter. Mankind had begun to really understand what was happening in the new, bizarre quantum world.

But while the crew of the Enterprise could always be teleported away from the most hostile worlds, the physicists of the early twentieth century did not go back: they realized that the strange laws they were discovering were fundamental and were the basis of the behavior of all matter in the universe. Since everything, including humans, is made of atoms, it is impossible to escape the consequences of what happens at the atomic level. We have discovered an alien world, and that world is within us!

The shocking consequences of their discoveries upset not a few scientists of the time. A bit like revolutionary ideologies, quantum physics consumed many of its prophets. In this case the ruin did not come from political machinations or conspiracies of adversaries, but from disconcerting and deep philosophical problems that had to do with the idea of reality. When, towards the end of the 1920s, it became clear to everyone that a real revolution had occurred in physics, many of those who had given it the initial impetus, including a figure of the caliber of Albert Einstein, repented and turned their backs on the theory they had contributed significantly to creating. Yet today, well underway in the 21st century, we use quantum physics and apply it to a thousand situations. Thanks to her we have invented for example transistors, lasers, atomic energy and countless other things. Some physicists, even prominent ones, continue to use all their strength to find a version of quantum mechanics softer for our common sense, less destructive than the common idea of reality. But it would be good to reckon with science, not with some palliative.

Before the quantum era, physics had managed very well to describe the phenomena that happen before our eyes, solving problems in a world made of stairs firmly resting on the walls, arrows and cannonballs launched according to precise trajectories, planets orbiting and rotating on themselves, comets returning to the expected time, steam engines doing their useful work, telegraphs and electric motors. In short, at the beginning of the twentieth century almost every observable and measurable macroscopic phenomenon had found a coherent explanation within the so-called classical physics. But the attempt to apply the same laws to the strange microscopic world of atoms proved incredibly difficult, with deep philosophical implications. The theory that seemed to come out, the quantum theory, went completely against common sense.

Our intuition is based on previous experiences, so we can say that even classical science, in this sense, was sometimes counterintuitive, at least for the people of the time. When Galileo discovered the laws of ideal motion in the absence of friction, his ideas were considered extremely daring (in a world where no one or almost no one had thought to neglect the effects of friction).2 But the classical physics that emerged from his intuitions managed to redefine common sense for three centuries, until the 20th century. It seemed a solid theory, resistant to radical changes - until quantum physics burst onto the scene, leading to an existential shock like never before.

To really understand the behavior of atoms, to create a theory that would agree with the seemingly contradictory data that came out of the laboratories in the thirty years between 1900 and 1930, it was necessary to act in a radical way, with a new audacity. The equations, which until then calculated with precision the dynamics of events, became tools to obtain fans of possibilities, each of which could happen with a given probability. Newton's laws, with their certainties (so we speak of "classical determinism") were replaced by Schrödinger's equations and Heisenberg's disconcerting mathematical constructions, which spoke the language of indeterminacy and nuance.

How does this uncertainty manifest itself in nature, at the atomic level? In various areas, of which here we can give a first, simple example. Atomic physics tells us that given a certain amount of radioactive material, let's say uranium, half will transform thanks to a process called "decay" and will disappear before a fixed period of time, called "half-life" or "half-life". After another time interval equal to the half-life, the remaining atoms will be reduced another time by half (so after a time as long as two half-lives the amount of uranium present at the beginning will be reduced to a quarter of the original; after three half-lives, to an eighth; and so on). Thanks to quantum mechanics and some complicated equations, we are able to calculate in principle the value of the half-life of uranium, and of many other fundamental particles. We can put at work arrays of theoretical physicists and get many interesting results. Yet, we are absolutely not able to predict when a certain uranium atom will decay.

It is a disconcerting result. If uranium atoms were to follow the laws of classical Newtonian physics, there would be some mechanism at work that, provided we perform the calculations accurately, would allow us to predict exactly when a certain atom will decay. Quantum laws do not offer deterministic mechanisms and give us probability and data blurred not for simple ignorance of the problem, but for deeper reasons: according to the theory, the probability that the decay of that atom happens in a certain period is all we can know.

Let's go to another example. Let's consider two identical photons (the particles of which the light is made) and shoot them in the direction of a window. There are several alternatives: both of them bounce on the glass, both of them cross it, one bounces and one crosses it. Well, quantum physics is not able to predict how the single photons will behave, whose future is not known even in principle. We can only calculate the probability with which the various alternatives will happen - for example that such a photon will be rejected at 10% and will pass to 90%, but nothing more. Quantum physics may seem vague and inaccurate at this point, but it actually provides the correct procedures (the only correct procedures, to be precise) that allow us to understand how matter works. It is also the only way to understand the atomic world, the structure and behavior of particles, the formation of molecules, the mechanism of radiation (the light we see comes from atoms). Thanks to her we were able, in a second time, to penetrate into the nucleus, to understand how the quarks that form protons and neutrons are bound together, how the Sun gets its gigantic energy, and more.

But how is it possible that the physics of Galileo and Newton, so tragically inadequate to describe the atomic motions, is able to predict with few, elegant equations the motions of celestial bodies, phenomena such as eclipses or the return of Halley's comet in 2061 (a Thursday afternoon) and the trajectories of spacecraft? It is thanks to classical physics that we can design the wings of airplanes, skyscrapers and bridges able to withstand strong winds and earthquakes, or robots able to perform high-precision surgery. Why does everything work so well, if quantum mechanics shows us with great evidence that the world does not work at all as we thought?

This happens: when huge amounts of atoms join together to form macroscopic objects, as in the examples we just did (airplanes, bridges and robots), the disturbing and counterintuitive quantum phenomena, with their load of uncertainty, seem to erase each other and bring the phenomena back to the bedrock of the precise predictability of Newtonian physics. The reason why this happens, in money, is of a statistical nature. When we read that the average number of members of American families is equal to 2,637 individuals we are faced with a precise and deterministic data. Too bad however that no family has exactly 2,637 members.

In the 21st century quantum mechanics has become the backbone of all the research in the atomic and subatomic world, as well as of wide sectors of material sciences and cosmology. The fruits of the new physics make thousands of billions of dollars every year, thanks to the electronics industry, and as many follow from the improvements in efficiency and productivity made possible by the systematic use of quantum laws. Some physicists a bit rebellious, however, driven by the cheers of a certain type of philosophers, still continues to seek a deeper meaning, a principle hidden within quantum mechanics in which determinism is found. But it is a minority.

Why is quantum physics disturbing, from a psychological point of view?

In a famous passage of a letter to Max Born, Einstein wrote: "You believe that God plays dice with the world, I believe instead that everything obeys a law, in a world of objective reality that I try to grasp through furiously speculative [...] Not even the great initial success of quantum theory manages to convince me that at the basis of everything there is randomness, although I know that younger colleagues consider this attitude as an effect of sclerosis.3 Erwin Schrödinger thought in a similar way: "If I had known that my wave equation would be used in this way, I would have burned the article before publishing it [...] I don't like it and I regret having had anything to do with it".4 What disturbed these eminent figures, so much so that they were forced to deny their beautiful creation? Let us go into a little detail about these lamentations, in Einstein's protest against a God who "plays dice". The turning point of modern quantum theory dates back to 1925, precisely to the solitary vacation that the young German physicist Werner Heisenberg spent in Helgoland, a small island in the North Sea where he had retired to find relief from hay fever. There he had a revolutionary idea.5

In the scientific community there was more and more support for the hypothesis that atoms were composed of a denser central nucleus surrounded by a cloud of electrons, similar to planets orbiting the Sun. Heisenberg examined the behavior of these electrons and realized that for his calculations it was not necessary to know their precise trajectories around the nucleus. The particles seemed to jump mysteriously from one orbit to another and at each jump the atoms emitted light of a certain color (the colors reflect the frequency of the light waves). From a mathematical point of view, Heisenberg had been able to find a sensible description of these phenomena, which however implied a different atom model from that of a tiny solar system, with planets confined on immutable orbits. In the end he forgot the calculation of the trajectory of an electron moving from the observed position A to B, because he realized that any measure of the particle in that time would necessarily interfere with its behavior. So Heisenberg elaborated a theory that accounted for the colors of the emitted light, without requiring the knowledge of the precise trajectory followed by the electron. In the end it counted only that a certain event was possible and happened with a certain probability. The uncertainty became an intrinsic feature of the system: the new reality of quantum physics was born.

Heisenberg's revolutionary solution to the problems posed by a series of disconcerting experimental data unleashed the fantasy of his mentor, Niels Bohr, father, grandfather and obstetrician of the new theory. Bohr took the young colleague's ideas to the extreme, so much so that Heisenberg himself was initially disturbed. In the end he changed his mind and converted to the new verb, which many of his eminent colleagues refused to do. Bohr had reasoned in this way: if to know which path has traveled a certain electron is not relevant for the calculation of atomic phenomena, then the very idea of "orbit", of an established trajectory like that of a planet around a star, must be abandoned because it is meaningless. Everything is reduced to observation and measurement: the act of measuring forces the system to choose between the various possibilities. In other words, it is not the uncertainty of measurement that hides reality; on the contrary, it is reality itself that never provides certainty in the classical-Galileian sense of the term, when examining phenomena on an atomic scale.

In quantum physics there seems to be a magical link between the physical state of a system and its conscious perception by a sentient observer. But it is the very act of measuring, that is the arrival on the scene of another system, that resets all but one possibility, making the quantum state "collapse", as they say, in one of the many alternatives. We will see how much this can be disturbing later on, when we will meet electrons that are passed one at a time through two slits in a screen and that form configurations that depend on the knowledge of the precise slit from which they passed, that is if someone or something has made a measurement on the system. It seems that a single electron, as if by magic, crosses the two slits at the same time if nobody is watching it, while choosing a possible path if someone or something is watching it! This is possible because electrons are neither particles nor waves: they are something else, completely new. They have been quantum.6

It is little wonder to discover that many of the pioneers of the new physics, who had participated in the creation of atomic science, were reluctant to accept these strange consequences. The best way to gild the pill and get Heisenberg and Bohr's theses accepted is the so-called "Copenhagen interpretation. According to this version of the facts, when we measure an atomic scale system we introduce in the system itself an important interference, given by the measuring instruments. But whatever interpretation we give, quantum physics does not correspond to our intuitive ideas of reality. We must learn to live with it, to play with it, to verify its goodness with experiments, to imagine theoretical problems that exemplify various situations, to make it more and more familiar. In this way we could develop a new "quantum intuition", however contrary to common sense it may seem at first.

In 1925, completely independent from Heisenberg's ideas, another theoretical physicist had another fundamental idea, also while he was on vacation (not alone, however). It was the Viennese Erwin Schrödinger, who had formed a bond of friendship and scientific collaboration with his colleague Hermann Weyl. The latter was a mathematician of great value, who had a decisive role in the development of the relativity theory and in the relativistic version of the electron theory. Weyl helped Schrödinger with the calculations and as compensation he could sleep with his wife Anny. We do not know what the woman thought about the matter, but social experiments of this kind were not rare in the twilight of the Viennese intellectual society. This agreement also included the possibility for Schrödinger to embark on a thousand extramarital adventures, one of which led (in a certain sense) to a great discovery in the quantum field.7

In December 1925, Schrödinger went on a twenty-day vacation to Arosa, a village in the Swiss Alps. Leaving Anny at home, he was accompanied by an old Viennese flame. In her suitcase she also put a scientific article by her French colleague Louis de Broglie and earplugs. While he was concentrating on the writing, sheltered from annoying noises (and who knows what the lady was doing in the meantime), the idea of the so-called "wave mechanics" came to his mind. It was a new and different way to formalize the emerging quantum theory in mathematically simpler terms, thanks to equations that were well known to the main physicists of the time. This revolutionary idea was of great support for the then fragile quantum theory, which became known to a much greater number of people.8 The new equation, which in honor of its discoverer is called "Schrödinger's equation", on one hand accelerated the path of quantum mechanics, but on the other hand made its inventor crazy because of the way it was interpreted. It is surprising to read of Schrödinger's repentance, due to the scientific and philosophical revolution triggered by his ideas.

The idea was this: to describe the electron with the mathematical tools used for the waves. This particle, that before was thought to be modelable as a microscopic ball, in some occasions it behaves just like a wave. The wave physics (phenomena found in many fields, from water to sound, from light to radio etc.) was well known at that time. Schrödinger was convinced that a particle like electron was really a new kind of wave, a "matter wave", so to speak. It seemed a bisque hypothesis, but the resulting equation was useful in calculations and gave concrete results in a relatively simple way. Schrödinger's wave mechanics gave comfort to those sectors of the scientific community whose members had great difficulty in understanding the apparently unstoppable quantum theory and who found Heisenberg's version too abstract for their tastes.

The central point of Schrödinger's idea is the type of solution of the equation that describes the wave. It is written by convention with the Greek capital letter psi, Ψ - the so-called "wave function". Ψ is a function in space and time variables that contains all information about the electron. Schrödinger's equation, therefore, tells us how the wave function varies as space and time changes.

Applied to the hydrogen atom, Schrödinger's equation made it possible to discover the behavior of the electron around the nucleus. The electronic waves determined by Ψ looked like the sound waves produced by a bell or some other musical instrument. It is like plucking the strings of a violin or guitar: the result is vibrations that correspond in a precise and observable way to various energy levels. Schrödinger equation gave the correct values of these levels corresponding to the electron oscillations. The data in the hydrogen atom case had already been determined by Bohr in his first attempt of theoretical arrangement (that today is called with some sufficiency "old quantum theory"). The atom emits light with well defined energy levels (the so called "spectral lines") that thanks to the quantum mechanics today we know are connected to the electron jumps, that passes from a state of motion associated to the wave say Ψ_2 to the one associated to the wave Ψ_1.

Schrödinger's equation proved to be a powerful tool, thanks to which to determine wave functions through purely mathematical methods. The same idea could be applied not only to electrons, but to any phenomenon that required a quantum level treatment: systems consisting of various electrons, whole atoms, molecules, crystals, conducting metals, protons and neutrons in the nucleus. Today we have extended the method to all particles composed by quarks, the fundamental bricks of nuclear matter.

For Schrödinger, electrons were pure and simple waves, similar to marine or sound waves, and their particle nature could be neglected as illusory. Ψ represented waves of new type, those of matter. But in the end this interpretation turned out to be wrong. What was Ψ really about? After all, electrons continued to behave as if they were point particles, which for example you could see when they collided with a fluorescent screen. How was this behavior reconciled with the undulatory nature?

Another German physicist, Max Born (who by the way was an ancestor of the singer Olivia Newton-John), proposed a new interpretation of Schrödinger's equation that remains today a cornerstone of physics. According to him, the wave associated to the electron was a so-called "probability wave".10 To be precise, the square of $\Psi(x, t)$, that is $\Psi_2(x, t)$, was the probability to find the electron in the x point at time t. Where the value of Ψ_2 is high, there is a strong probability to find the electron. Where

$\Psi_2=0$, instead, there is no possibility. It was a shocking proposal, similar to Heisenberg's one, but it had the merit to be easier to understand, because it was formulated in the most familiar ground of Schrödinger equation. Almost everyone was convinced and the question seemed to be closed.

Born's hypothesis clearly states that we do not know and we will never be able to know where the electron is. Is it maybe there? Mah, there is an 85 percent probability that it is so. Is it on the other side? We cannot exclude it, there is a 15% probability. Born's interpretation defines also without hesitation what can or cannot be predicted in the experiments, and it does not exclude the case that two apparently identical tests give very different results. It seems that particles can afford the luxury of being where they are at a certain time without obeying those strict rules of causal type that are usually associated with classical physics. There is nothing to do, for quantum theory it seems that God is playing dice with the universe.

Schrödinger was not happy to have been a protagonist of that disturbing revolution. Together with Einstein, who ironically wrote an article in 1911 that gave Born the inspiration for his idea, he remained in the field of dissidents all his life. Another "passerby" was the great Max Planck, who wrote: "The probabilistic interpretation proposed by the Copenhagen group must be condemned without fault, for high treason against our beloved physicist".11

Planck was one of the greatest theoretical physicists active at the turn of the century, and even he did not like the fold that quantum theory had taken. It was the supreme paradox, since he had been the true founder of the new physics, besides having coined the term "how much" already at the end of the nineteenth century.

We can perhaps understand the scientist that speaks of "betrayal" about the entry of the probability in the physical laws instead of the solid certainties of cause and effect. Let's imagine having a normal tennis ball and bouncing it against a smooth concrete wall. We do not move from the point where we threw it and continue to hit it with the same force and aiming in the same direction. Under the same boundary conditions (such as wind), a good tennis player should be able to get the ball in exactly the same place, shot after shot, until he gets tired or the ball (or the wall) breaks. A champion like Andre Agassi counted on these characteristics of the physical world to develop in training the skills that allowed him to win Wimbledon. But what would happen if the rebound was not predictable? Or if even on some occasion the ball was able to cross the wall? What if only the probability of the phenomenon is known? For example, fifty-five times out of a hundred the ball goes back, the other forty-five times it goes through the wall. And so on, for everything: there is also a probability that it will cross the barrier formed by the racket. We know that this never happens in the macroscopic and Newtonian world of tennis tournaments. But at the atomic level everything changes. An electron shot against the equivalent of a particle wall has a probability different from zero to cross it, thanks to a property known as "tunnel effect". Imagine what kind of difficulties and frustrations would meet a tennis player engaged in the subatomic world.

There are cases, however, where the non-deterministic behavior is observable in everyday reality, especially that of photons. You are looking at the window of a store window full of sexy underwear, and you realize that a faded image of yourself has formed on the dummy's shoes. Why is that? The phenomenon is due to the nature of light, a stream of particles (photons, in fact) with bizarre quantum properties. The photons, which we suppose come from the Sun, mostly bounce on your face, go through the glass and show a clear image of you (but you are not bad) to the person inside the store (maybe the window dresser who is dressing the mannequin). But a small part of the photons is reflected by the glass and gives your eyes the faint portrait of your face lost in the contemplation of those microscopic clothes. But how is this possible, since all the photons are identical?

Even with the most sophisticated experiments, there is no way to predict what will happen to the photons. We are given to know only the probability of the event: applying Schrödinger's equation, we can calculate that the luminous particles pass through the window 96 times out of 100 and bounce back the remaining 4 times. Are we able to know what the single photon does? No, in no way, not even with the best tools imaginable. God rolls the dice every time to decide where to pass the particle, or so quantum physics tells us (maybe he prefers roulette... anyway, it is clear that he plays with probabilities).

To replicate the showcase situation in an experimental context (and much more expensive), we shoot electrons against a barrier constituted by a grid of conductive wires inside a container where the vacuum is made, connected to the negative pole of a battery with voltage equal, for example, to 10 volts. An electron with energy equivalent to a potential of 9 volt should be reflected, because it cannot counter the repulsive force of the barrier. But Schrödinger's equation tells us that a part of the wave associated to the electron is still able to pass through it, just as it happened to photons with glass. But in our experience there are no "fractions" of photon or electron: these particles are not made of plasticine and you cannot detach pieces of them at will. So the final result must always be only one, that is reflection or crossing. If the calculations tell us that the first eventuality occurs in 20 percent of the cases, this means that all the electron or photon is reflected with 20 percent probability. We know this thanks to Schrödinger's equation, which gives us the result in terms of $\Psi 2$.

It was with the help of analogous experiments that physicists abandoned the original Schrödinger's interpretation, that was to say that it provided "plasticine" electrons, i.e. matter waves, to arrive to the probabilistic one, much less intuitive, according to which a certain mathematical function, $\Psi 2$, provided the probability to find the particles in a certain position at a given instant. If we shoot a thousand electrons against a screen and check with a Geiger counter how many of them pass it, we find maybe that 568 have passed and 432 have been reflected. Which of them was this fate? There is no way to know, neither now nor ever. This is the frustrating reality of quantum physics. All we can do is calculate the probability of the event, $\Psi 2$.

Schrödinger had a kitten

In examining the philosophical paradoxes brought by quantum theory we cannot overlook the now famous case of Schrödinger's cat, in which the funny microscopic world with its probabilistic laws is linked to the macroscopic one with its precise Newtonian pronouncements. Like Einstein, Podolsky and Rosen, Schrödinger did not want to accept the fact that objective reality did not exist before observation, but was in a jumble of possible states. His cat paradox was originally intended as a way of mocking a worldview that was untenable for him, but it has proved to be one of the most tenacious nightmares of modern science to this day. This too, like EPR, is a mental or conceptual experiment, designed to make quantum effects resoundingly manifest even in the macroscopic field. And also this makes use of radioactivity, a phenomenon that involves the decay of matter according to a predictable rate, but without knowing exactly when the single particle will disintegrate (that is, as we have seen above, we can say how many particles will decay in an hour, for example, but not when one of them will).

Here is the situation imagined by Schrödinger. We close a cat inside a box together with a vial containing a poisonous gas. On the other hand, we put a small well sealed amount of radioactive material, so that we have a 50% chance of observing a single decay in the space of one hour. Let's invent some kind of device that connects the Geiger counter that detects the decay to a switch, which in turn activates a hammer, which in turn hits the vial, thus releasing the gas and killing the cat (of course these Viennese intellectuals of the early twentieth century were very strange...).

Let an hour pass and ask ourselves: is the cat alive or dead? If we describe the system with a wave function, we get a "mixed" state15 like the one seen above, in which the cat is "smeared" (we apologize to cat lovers) in equal parts between life and death. In symbols we could write Ψgatto-vivo + Ψgatto-morto. At macroscopic level, we can only calculate the probability of finding the cat alive, equal to (Ψgatto-vivo)2, and that of finding it dead, equal to (Ψgatto-morto)2.

But here is the dilemma: the collapse of the initial quantum state in the "live cat" or "dead cat" is determined by the moment someone (or something) peeks inside the box? Couldn't the cat itself, who observes the Geiger counter distressed, be the entity capable of the measurement? Or, if we want a deeper identity crisis: the radioactive decay could be monitored by a computer, which at any time is able to print the state of the cat on a sheet of paper inside the box. When the computer records the arrival of the particle, is the cat definitely alive or dead? or is it when the printing of the state is finished? or when a human observer reads it? or when the flow of electrons produced by the decay meets a sensor inside the Geiger counter that triggers it, that is when we move from the subatomic world to the macroscopic one? Schrödinger's cat paradox, like the EPR one, seems at first sight a strong refutation of the fundamental principles of quantum physics. It is clear that the cat cannot be in a "mixed" state, half alive and half dead. Or not?

As we will see better later, some experiments have shown that Schrödinger's visible cat, representative of all macroscopic systems, can really be in a mixed state; in other words, quantum theory implies the existence of these situations also at macroscopic level. Another victory for the new physics.

Quantum effects, indeed, can occur at various scales, from the smallest of atoms to the largest of systems. An example is given by the so-called "superconductivity", so that at very low temperatures certain materials have no electrical resistance and allow the current to circulate infinitely without the help of batteries, and the magnets to remain suspended above the circuits forever. The same goes for "superfluidity", a state of matter in which, for example, a flow of liquid helium can rise up the walls of a test tube or feed perpetual fountains, without consuming energy. And the same is also true for the mysterious phenomenon thanks to which all particles acquire mass, the so-called "Higgs mechanism". There is no way to escape quantum mechanics: in the end, we are all cats locked in some box.

No math, promised, but just a few numbers

With this book we would like to give you an idea of the tools that physics has developed to try to understand the strange microscopic world inhabited by atoms and molecules. We ask readers only two small efforts: to have a healthy sense of curiosity about the world and to master the advanced techniques for solving differential equations with partial derivatives. All right, we joked. After years of giving elementary physics courses to students of non-scientific faculties, we know how widespread the terror of mathematics is among the population. No formulas, then, or at least the minimum necessary, few and scattered here and there.

The scientific vision of the world should be taught to everyone. Quantum mechanics, in particular, is the most radical change of perspective occurred in human thought since the ancient Greeks began to abandon the myth in favor of the search for rational principles in the universe. Thanks to the new theory, our understanding of the world has greatly expanded. The price paid by modern science for this broadening of intellectual horizons has been the acceptance of many apparently counterintuitive ideas. But remember that the blame for this falls mainly on the shoulders of our old Newtonian language, unable to accurately describe the atomic world. As scientists, we promise to do our best.

Since we are about to enter the realm of the infinitely small, it is convenient for us to use the convenient notation of the "powers of ten". Do not be frightened by this scientific shorthand that we will sometimes use in the book: it is just a method to effortlessly record very large or very small numbers. If you see written for example 104 ("ten high to the fourth power", or "ten to the fourth"), all you have to do is translate it "one followed by four zeros": 104=10000. On the contrary, 10-4 indicates "one preceded by four zeros", one of which must obviously be before the comma: 10-4=0.0001, that is 1/10000, one ten thousandth.

Using this simple language, let's see how to express the scales at which various natural phenomena occur, in descending order.

- 100 m=1 m, that is one meter: it is the typical human scale, equal to the height of a child, the length of an arm or a step;

- 10-2 m=1 cm, that is one centimeter: it is the width of an inch, the length of a bee or a hazelnut.

- 10-4 m, a tenth of a millimeter: is the thickness of a pin or the legs of an ant; so far we are always in the domain of application of classical Newtonian physics.

- 10-6 m, one micron or millionth of a meter: we are at the level of the largest molecules found in the cells of organisms, such as DNA; we are also at the wavelength of visible light; here we begin to feel the quantum effects.

- 10^{-9} m, one nanometer or billionth of a meter: it is the diameter of a golden atom; the smallest among the atoms, the hydrogen one, has a diameter of 10^{-10} m.

- 10^{-15} m: we are in the parts of atomic nucleus; protons and neutrons have a diameter of 10^{-16} m, and under this length we find quarks.

- 10^{-19} m: it is the smallest scale that can be observed with the most powerful particle accelerator in the world, the LHC of CERN in Geneva.

- 10^{-35} m: is the smallest scale that we believe exists, under which the very idea of "distance" loses its meaning due to quantum effects.

The experimental data tell us that quantum mechanics is valid and fundamental for the understanding of phenomena from 10^{-9} to 10^{-15} meters, i.e. from atoms to nuclei (in words: from one billionth to one millionth of one billionth of a meter). In some recent researches, thanks to the Tevatron of Fermilab, we have been able to probe distances of the order of 10^{-18} meters and we have not seen anything that convinced us of the failure at that scale of quantum mechanics. Soon we will penetrate into territories smaller by a factor of ten, thanks to the colossal LHC, the CERN accelerator that is about to start working.* The exploration of these new worlds is not similar to the geographical one, to the discovery of a new continent previously unknown. It is rather an investigation within our world, because the universe is formed by the collection of all the inhabitants of the microscopic domain. On its properties, and their consequences, our future depends.

Why do we need a "theory"?

Some of you may ask yourselves if it is worth the effort for a simple theory. Well, there are theories and theories, and it is the fault of us scientists who use the same word to indicate very different contexts. In itself, a "theory" is not even well defined from a scientific point of view.

Let's make an example a bit banal. A population living on the shores of the Atlantic Ocean notices that the Sun rises on the horizon every morning at 5 a.m. and sets in the opposite direction every night at 7 p.m. To explain this phenomenon, a venerable wise man puts forward a theory: there are an infinite number of suns hidden under the horizon, which appear every 24 hours. There is, however, a hypothesis that requires less resources: only one Sun revolving around the Earth, supposedly spherical, in 24 hours. A third theory, the most bizarre and counterintuitive, argues instead that the Sun stands still and the Earth turns on itself in 24 hours. We have therefore three ideas in conflict between them. In this case the word "theory" implies the presence of a hypothesis that explains in a rational and systematic way why what we observe happens.

The first theory is easily refuted, for many good reasons (or simply because it is idiotic). More difficult is to get rid of the second one; for example you could observe that the other planets in the sky rotate on themselves, so by analogy the Earth should do it too. Be that as it may, in the end thanks to accurate experimental measurements we verify that it is our world that is rotating. So only one theory survives, which we will call axial rotation or RA.

Here is a problem, however: in all the previous discussion we have never talked about "truth" or "facts", but only "theories". We know very well that the RA is centuries old, and yet we still call it "Copernican theory", even if we are sure that it is true, that it is an established fact. In reality we want to emphasize the fact that the RA hypothesis is the best, in the sense that it better fits with the observations and tests, very different and even performed in extreme circumstances. Until we have a better explanation, we keep this one. Yet we continue to call it theory. Perhaps because we have seen in the past that ideas taken for granted in some areas have required changes in the transition to different areas.

So today we talk about "theory of relativity", "quantum theory", "theory of electromagnetism", "Darwinian theory of evolution" and so on, even if we know that all these have reached a higher degree of credibility and scientific acceptance. Their explanations of the various phenomena are valid and are considered as "factual truths" in their respective domains of application. There are also theories proposed but not verified, such as that of strings, which seem excellent attempts, but could be accepted as rejected. And there are theories definitively abandoned, like that of the phlogiston (a mysterious fluid responsible for combustion) and the caloric (an equally mysterious fluid responsible for heat transmission). Today, however, quantum theory is the best verified among all the scientific theories ever proposed. Let's accept it as a fact: it is a fact.

Enough with the intuitive, hurray for the counterintuitive

As we approach the new atomic territories, everything that intuition suggests becomes suspicious and the information accumulated so far may no longer be useful to us. Everyday life takes place within a very limited range of experiences. We do not know, for example, what it feels like to travel a million times faster than a bullet, or to endure temperatures of billions of degrees; nor have we ever danced in the full moon with an atom or a nucleus. Science, however, has compensated for our limited direct experience of nature and made us aware of how big and full of different things the world out there is. To use the metaphor dear to a colleague of ours, we are like chicken embryos that feed on what they find in the egg until the food is finished, and it seems that our world must end too; but at that point we try to give the shell a shot, we go out and discover an immensely larger and more interesting universe.

Among the various intuitions typical of an adult human being there is the one that the objects that surround us, whether they are chairs, lamps or cats, exist independently from us and have certain objective properties. We also believe, based on what we study in school, that if we repeat an experiment at various times (for example if we let two different cars run along a ramp) we should always get the same results. It is also obvious, intuitive, that a tennis ball passing from one half of the court to the other has a defined position and speed at all times. It is enough to film the event, i.e. obtain a collection of snapshots, to know the situation at various moments - and to reconstruct the overall trajectory of the ball.

These intuitions continue to help us in the macroscopic world, between machines and balls, but as we have already seen (and we will see again in the course of the book), if we go down to the atomic level we see strange things happening, which force us to abandon the preconceptions dear to us: prepare to leave your intuitions at the entrance, dear readers. The history of science is a history of revolutions, but they do not throw away all previous knowledge. Newton's work, for example, understood and extended (without destroying them) the previous research of Galileo, Kepler and Copernicus. James Clerk Maxwell, inventor of the theory of electromagnetism in the nineteenth century,[17] took the results of Newton and used them to extend certain aspects of the theory to other fields. The Einsteinian relativity incorporated Newton's physics and extended the domain until to understand cases in which the speed is very high or the space very extended, fields in which the new equations are valid (while the old ones remain valid in the other cases). Quantum mechanics started from Newton and Maxwell to arrive to a coherent theory of atomic phenomena. In all these cases, the passage to the new theories was done, at least at the beginning, using the language of the old ones; but with quantum mechanics we see the failure of the classical language of the previous physics, as well as of the natural human languages.

Einstein and his dissident colleagues were faced with our own difficulty, that is, to understand the new atomic physics through the vocabulary and philosophy of macroscopic objects. We have to learn to understand that the world of Newton and Maxwell finds itself as a consequence of the new theory, which is expressed in the quantum language. If we were also as big as atoms, we would have grown up surrounded by phenomena that would be familiar to us; and maybe one day an alien as big as a quark would ask us: "What kind of world do you think we get if we put together 1023 atoms and form an object that I call "ball"?

Perhaps it is the concepts of probability and indeterminacy that challenge our linguistic abilities. This is no small problem that remains in our day and frustrates even great minds. It is said that the famous theoretical physicist Richard Feynman refused to answer a journalist who, during an interview, asked him to explain to the public what force was acting between two magnets, claiming that it was an impossible task. Later, when asked for clarification, he said it was because of intuitive preconceptions. The journalist and a large part of the audience understand "force" as what we feel if we reward the palm of your hand against the table. This is their world, and their language. But in reality in the act of supporting the hand are involved electromagnetic forces, the cohesion of matter, quantum mechanics - it is very complicated. It was not possible to explain the pure magnetic force in familiar terms to the inhabitants of the "old world".

As we will see, to understand quantum theory we must enter a new world. It is certainly the most important fruit of the scientific explorations of the twentieth century, and it will be essential for the whole new century. It is not right to let only professionals enjoy it.

Even today, at the beginning of the second decade of the 21st century, some illustrious scientists continue to search with great effort for a more "friendly" version of quantum mechanics that less disturb our common sense. But these efforts so far do not seem to lead to anything. Other scientists simply learn the rules of the quantum world and make progress, even important ones, for example adapting them to new principles of symmetry, using them to hypothesize a world where strings and membranes replace elementary particles, or imagining what happens at scales billions of times smaller than those we have reached so far with our instruments. This last line of research seems the most promising and could give us an idea of what could unify the various forces and the very structure of space and time.

Our aim is to make you appreciate the disturbing strangeness of quantum theory, but above all the profound consequences it has on our understanding of the world. For our part, we think that the uneasiness is mainly due to our prejudices. Nature speaks in a different language, which we must learn - just as it would be good to read Camus in the original French and not in a translation full of American slang. If a few steps gives us a hard time, let's take a nice vacation in Provence and breathe the air of France, rather than stay in our house in the suburbs and try to adapt the language we use every day to that very different world. In the next chapters we will try to transport you to a place that is part of our universe and at the same time goes beyond imagination, and in the next chapters we will also teach you the language to understand the new world.

Chapter 2
The beginning

A complicating factor

Before trying to understand the dizzying quantum universe, it is necessary to familiarize oneself with some aspects of the scientific theories that preceded it, that is, with the so-called classical physics. This body of knowledge is the culmination of centuries of research, begun even before Galilco's time and completed by geniuses such as Isaac Newton, Michael Faraday, James Clerk Maxwell, Heinrich Hertz and many others.2 Classical physics, which reigned unchallenged until the early 20th century, is based on the idea of a clockwork universe: ordered, predictable, governed by causal laws.

To have an example of a counterintuitive idea, let us take our Earth, which from our typical point of view appears solid, immutable, eternal. We are able to balance a tray full of cups of coffee without spilling a drop, yet our planet spins fast on itself. All the objects on its surface, far from being at rest, spin with it like the passengers of a colossal carousel. At the Equator, the Earth speeds faster than a jet, at over 1600 kilometers per hour;

moreover, it runs wildly around the Sun at an incredible average speed of 108,000 kilometers per hour. And on top of that, the entire solar system, including the Earth, travels the galaxy at even higher speeds. Yet we do not realize it, we do not feel like we are running. We see the Sun rising in the east and setting in the west, and nothing more. How is this possible? Writing a letter while riding a horse or driving a car at 100 km/h on the highway is a very difficult task, yet we have all seen footage of astronauts doing precision work inside an orbiting station, launched around our planet at almost 30,000 kilometers per hour. If it weren't for the blue globe changing shape in the background, those men floating in space seem to be standing still.

Intuition, in general, does not notice if what surrounds us is moving at the same speed as us, and if the motion is uniform and not accelerated we do not feel any sensation of displacement. The Greeks believed that there was a state of absolute rest, relative to the surface of the Earth. Galileo questioned this venerable Aristotelian idea and replaced it with a more scientific one: for physics there is no difference between standing still and moving with constant (even approximate) direction and speed. From their point of view, astronauts are standing still; seen from Earth, they are circling us at a crazy speed of 28 800 kilometers per hour.

Galileo's sharpened ingenuity easily understood that two bodies of different weights fall at the same speed and reach the ground at the same time. For almost all his contemporaries, however, it was anything but obvious, because daily experience seemed to say otherwise. But the scientist did the right experiments to prove his thesis and also found a rational justification: it was the resistance of the air that shuffled the cards. For Galileo this was only a complicating factor, which hid the deep underlying simplicity of natural laws. Without air between the feet, all bodies fall with the same speed, from the feather to the colossal rock.

It was then discovered that the gravitational attraction of the Earth, which is a force, depends on the mass of the falling object, where mass is a measure of the amount of matter contained in the object itself.

The weight, instead, is the force exerted by gravity on bodies endowed with mass (perhaps you will remember that the physics teacher in high school repeated: "If you transport an object on the Moon, its mass remains the same, while the weight is reduced". Today all this is clear to us thanks to the work of men like Galileo). The force of gravity is directly proportional to the mass: you double the mass and the force is also doubled. At the same time, however, as the mass grows, so does the resistance to change the state of motion. These two equal and opposite effects cancel each other out and so it happens that all bodies fall to the ground with the same speed - as usual neglecting that complicating factor of friction.

To the philosophers of ancient Greece the state of rest seemed obviously the most natural for bodies, to which they all tend. If we kick a ball, sooner or later it stops; if we run out of fuel in a car, it stops too; the same happens to a disc slid on a table. All this is perfectly sensible and also perfectly Aristotelian (this of Aristotelianism must be our innate instinct).

But Galileo had deeper ideas. He realized, in fact, that by hinging the surface of the table and smoothing the puck, it would continue to run for a much longer time; we can verify this, for example, by sliding a field hockey puck over an icy lake. Let's remove all friction and other complicating factors, and see that the puck continues to slide infinitely along a straight trajectory at uniform speed. This is what causes the end of the motion, Galileo said: the friction between puck and table (or between car and road), that is a complicating factor.

Usually in the physics labs there is a long metal rail with numerous small holes through which air passes. In this way a trolley placed on the rail, the equivalent of our disk, can move floating on an air cushion. At the ends of the track there are rubber buffers. A small initial push is enough and the trolley starts bouncing non-stop between the two ends, back and forth, sometimes for the whole hour. It seems animated with its own life, how is it possible? The show is amusing because it goes against common sense, but in reality it is a manifestation of a profound principle of physics, which manifests itself when we remove the complication of friction. Thanks to less technological but equally enlightening experiments, Galileo discovered a new law of nature, which reads: "An isolated body in motion maintains its state of motion forever. By "isolated" we mean that no friction, various forces or anything else act on it. Only the application of a force can change a state of motion.

It is counterintuitive, isn't it? Yes, because it is very difficult to imagine a truly isolated body, a mythological creature that you do not encounter at home, in the park or anywhere else on Earth. We can only approach this ideal situation in the laboratory, with equipment designed as needed. But after witnessing some other version of the air track experiment, usually first year physics students end up taking the principle for granted.

The scientific method involves careful observation of the world. One of the cornerstones of its success in the last four centuries is its ability to create abstract models, to refer to an ideal universe in our mind, devoid of the complications of the real one, where we can hunt for the laws of nature. Having achieved a result in this world, we can go on the attack of the other, the more complicated one, after having quantified complication factors such as friction.

Let's move on to another important example. The solar system is actually intricate. There is a big star in the center, the Sun, and there are nine (or rather eight, after the downgrading of Pluto) smaller bodies of various masses around it; the planets in turn may have satellites. All these bodies attract each other and move according to a complex choreography. To simplify the situation, Newton reduced everything to an ideal model: a star and a single planet. How would these two bodies behave?

This research method is called "reductionist". Take a complex system (eight planets and the Sun) and consider a more treatable subset of it (one planet and the Sun). Now maybe the problem can be faced (in this case yes). Solve it and try to understand what characteristics of the solution are preserved in the return to the starting complex system (in this case we see that each planet behaves practically as if it were alone, with minimal corrections due to the attraction between the planets themselves).

The reductionism is not always applicable and does not always work. That is why we still do not have a precise description of objects like tornadoes or the turbulent flow of a fluid, not to mention the complex phenomena at the level of molecules and living organisms. The method proves useful when the ideal model does not deviate too much from its ugly and chaotic version, the one we live in. In the case of the solar system, the mass of the star is so much greater than that of the planets that it is possible to overlook the attraction of Mars, Venus, Jupiter and company when we study the motions of the Earth: the star + planet system provides an acceptable description of the Earth motions. And as we become familiar with this method, we can go back to the real world and make an extra effort to try to take into account the next complication factor in order of importance.

The parabola and the pendulum

Classical physics, or pre-quantistic physics, is based on two cornerstones. The first is the Galilean-Newtonian mechanics, invented in the seventeenth century. The second is given by the laws of electricity, magnetism and optics, discovered in the nineteenth century by a group of scientists whose names, who knows why, all remember some units of physical magnitude: Coulomb, Ørsted, Ohm, Ampère, Faraday and Maxwell. Let's start with Newton's masterpiece, the continuation of the work of our hero Galileo.

The bodies left free fall, with a speed that increases with the passage of time according to a fixed value (the rate of variation of the speed is called acceleration). A bullet, a tennis ball, a cannonball all describe in their motion an arc of supreme mathematical elegance, tracing a curve called parabola. A pendulum, that is, a body tied to a hanging wire (like a swing made by a tire tied to a branch, or an old clock), oscillates with a remarkable regularity, so that (precisely) you can adjust the clock. Sun and Moon attract the waters of the terrestrial seas and create tides. These and other phenomena can be explained rationally by Newton's laws of motion.

His creative explosion, which has few equals in the history of human thought, led him in a short time to two great discoveries. In order to describe them precisely and compare his predictions with data, he used a particular mathematical language called infinitesimal calculus, which for the most part he had to invent from scratch. The first discovery, usually labeled as "the three laws of motion", is used to calculate the motions of bodies once known the forces acting on them (Newton could have boasted so:

"Give me the forces and a computer powerful enough and I'll tell you what will happen in the future. But it seems he never said it).

The forces acting on a body can be exercised in a thousand ways: through ropes, sticks, human muscles, wind, water pressure, magnets and so on. A particular natural force, gravity, was at the center of Newton's second great discovery. Describing the phenomenon with an equation of astonishing simplicity, he established that all objects endowed with mass attract each other and that the value of the force of attraction decreases as the distance between the objects increases, in this way: if the distance doubles, the force is reduced by a quarter; if it triples, by a ninth; and so on. It is the famous "law of the inverse of the square", thanks to which we know that we can make the value of the force of gravity small at will, simply by moving away enough. For example, the attraction exerted on a human being by Alpha Centauri, one of the nearest stars (only four light years from here), is equal to one ten thousandth of a billionth, that is 10-13, of that exerted by the Earth. Vice versa, if we approach an object of great mass, like a neutron star, the resulting gravity force would crush us to the size of an atomic nucleus. Newton's laws describe the action of gravity on everything: apples falling from trees, bullets, pendulums and other objects located on the Earth's surface, where we almost all spend our existence. But they also apply in the vastness of space, for example between the Earth and the Sun, which are on average 150 million kilometers away.

Are we sure, however, that these laws are still valid outside our planet? A theory is valid if it provides values according to the experimental data (taking into account the inevitable measurement errors). Think a little bit: the evidence shows that Newton's laws work well in the solar system. With very good approximation, the single planets can be studied thanks to the simplification seen above, i.e. neglecting the effects of the others and taking into account the Sun and that's it. Newtonian theory predicts that the planets rotate around our star following perfectly elliptical orbits. But examining well the data we realize that there are small discrepancies in the case of Mars, whose orbit is not exactly the one provided by the "two bodies" approximation.

In studying the Sun-Mars system, we neglect the effects (relatively small) on the red planet of bodies like Earth, Venus, Jupiter and so on. The latter, in particular, is very large and gives some nice bang to Mars every time the orbits approach. In the long run, these effects add up. It is not excluded that in a few billion years Mars will be kicked out of the solar system as the competitor of a reality show. So we realize that the problem of planetary motions becomes more complex if we consider long time intervals. But thanks to modern computers we can face these small (and not so small) perturbations - including those due to Einstein's theory of general relativity, which is the modern version of Newtonian gravitation. With the right corrections, we see that the theory is always in perfect agreement with the experimental data. What can we say, however, when even greater distances come into play, such as those between the stars? The most modern astronomical measurements tell us that the force of gravity is present throughout the cosmos and for what we know it is valid everywhere.

Let's take a moment to contemplate a list of phenomena that take place according to Newtonian laws. Apples fall down from the trees, actually heading towards the center of the Earth. Artillery bullets sow destruction following parabola arches. The Moon looms only 384 000 kilometers from us and is the cause of tides and romantic languor. The planets whizz around the Sun along orbits that are not very elliptical, almost circular. Comets, on the other hand, follow very elliptical trajectories and take hundreds or thousands of years to make a turn and return to show themselves. From the smallest to the largest, the ingredients of the universe behave in perfectly predictable ways, following the laws discovered by Sir Isaac.

Chapter 3
What is light?

Before we leave classical physics behind us, we have to spend a few minutes talking about light and playing with it, because it will be the protagonist of many important (and at first puzzling) issues when we start to enter the quantum world. So we will now take a historical look at the theory of light in the classical world.[1]

Light is a form of energy. It can be produced in various ways, transforming electrical energy (as seen for example in a light bulb, or in the redness of toaster resistances) or chemical energy (as in candles and combustion processes in general). The sunlight, a consequence of the high temperatures present on the surface of our star, comes from nuclear fusion processes that take place inside it. And also the radioactive particles produced by a nuclear reactor here on Earth emit a blue light when they enter the water (which ionize, i.e. tear electrons from atoms).

It only takes a small amount of energy put into any substance to heat it. At small scales, this can be felt as a moderate temperature increase (as well knows also those who dabble with DIY on weekends, nails get warm after a series of hammering, or if they are torn from the wood with pliers). If we supply enough energy to a piece of iron, it begins to emit light radiation; initially it is reddish in color, then as the temperature increases we see orange, yellow, green and blue tones appear in order. At the end, if the heat is high enough, the emitted light becomes white, the result of the sum of all colors.

Most of the bodies around us, however, are visible not because they emit light, but because they reflect it. Excluding the case of mirrors, the reflection is always imperfect, not total: a red object appears to us as such because it reflects only this component of the light and absorbs orange, green, violet and so on. The pigments of paints are chemical substances that have the property of accurately reflecting certain colors, with a selective mechanism. White objects, on the other hand, reflect all the components of light, while black ones absorb them all: this is why the dark asphalt of a parking lot becomes hot on summer days, and this is the reason why in the tropics it is better to dress in light-colored clothing. These phenomena of absorption, reflection and heating, in relation to the various colors, have properties that can be measured and quantified by various scientific instruments.

Light is full of oddities. Here you are, we see you because the light rays reflected from your body affect our eyes. How interesting! Our mutual friend Edward is observing the piano instead: the rays of the you-us interaction (normally invisible, except when we are in a dusty or smoky room) intersect with those of the Edward-piano interaction without any apparent interference. But if we concentrate on an object the beams produced by two flashlights, we realize that the intensity of the lighting doubles, so there is interaction between the light rays.

Let us now examine the goldfish tank. We turn off the light in the room and turn on a flashlight. Helping ourselves with some dust suspended in the air, maybe produced by banging two blackboard erasers or a dust rag, we see that the light rays bend when they hit the water (and also that the poor little fish is looking at us perplexed, hopefully waiting for the feed). This phenomenon whereby transparent substances such as glass deflect light is called refraction. When the Boy Scouts light a fire by concentrating the sun's rays on a little dry wood through a lens, they are taking advantage of this property: the lens bends all the light rays making them concentrate in a point called "fire", and this increases the amount of energy until it is so high that it triggers the combustion.

A glass prism is able to decompose light into its components, the so-called "spectrum". These correspond to the colors of the rainbow: red, orange, yellow, green, blue, indigo and violet (to memorize the order remember the initials RAGVAIV). Our eyes react to this type of light, called "visible", but we know that there are also invisible types. On one side of the spectrum there is the so-called "infrared" long wave range (of this type, for example, is the radiation produced by certain heaters, the toaster resistances or the embers of a dying fire); on the other side there are the "ultraviolet" rays, short wave (an example is the radiation emitted by an arc welding machine, and that is why those who use it must wear protective glasses). The white light, therefore, is a mixture of various colors in equal parts. With special instruments we are able to quantify the characteristics of each color band, more properly their wavelength, and report the results on a graph. By subjecting any light source to this measurement, we find that the graph assumes a bell shape (see Fig. 4.1 below), whose peak is at a certain wavelength (i.e. color). At low temperatures the peak corresponds to long waves, i.e. red light. Increasing the heat, the maximum of the curve moves to the right, where the short waves are, i.e. violet light, but up to certain temperature values the amount of other colors is sufficient to ensure that the emitted light remains white. After these thresholds, the objects emit blue glow. If you look at the sky on a clear night, you will notice that the stars shine with slightly different colors: those tending to reddish are colder than white ones, which in turn are colder than blue ones. These gradations correspond to different stages of evolution in the life of the stars, as they consume their nuclear fuel. This simple identity card of light was the starting point of quantum theory, as we will see in more detail in a little while.

How fast does light travel?

The fact that light is an entity that "travels" in space, for example from a light bulb to our retina, is not entirely intuitive. In the eyes of a child, light is something that shines, not that moves. But that's just the way it is. Galileo was one of the first to try to measure its speed, with the help of two assistants placed on top of two nearby hills who spent the night covering and discovering two lanterns at predetermined times. When they saw the other light, they had to communicate it aloud to an external observer (Galileo himself), who made his measurements by moving at various distances from the two sources. It is an excellent way to measure the speed of sound, according to the same principle that a certain amount of time elapses between seeing lightning and hearing thunder. The sound is not very fast, it goes at about 1200 per hour (or 330 meters per second), so the effect is perceivable to the naked eye: for example, it takes 3 seconds before the thunderbolt comes from a lightning bolt that falls one kilometer away. But Galileo's simple experiment was not suitable to measure the speed of light, which is enormously higher.

In 1676 a Danish astronomer named Ole Römer, who at the time worked at the Paris Observatory, pointed his telescope towards the then known Jupiter satellites (called "Galileans" or "Medici" because they were discovered by the usual Galileo less than a century earlier and dedicated by him to Cosimo de' Medici). 2 He focused on their eclipses and noticed a delay with which the moons disappeared and reappeared behind the big planet; this small time interval depended mysteriously on the distance between Earth and Jupiter, which changes during the year (for example Ganymede seemed to be early in December and late in

July). Römer understood that the effect was due to the finite speed of light, according to a principle similar to that of the delay between thunder and lightning.

In 1685 the first reliable data on the distance between the two planets became available, which combined with the precise observations of Römer allowed to calculate the speed of light: it resulted an impressive value of 300000 kilometers per second, immensely greater than that of sound. In 1850 Armand Fizeau and Jean Foucault, two skilled French experimenters in fierce competition with each other, were the first to calculate this speed with direct methods, on Earth, without resorting to astronomical measurements. It was the beginning of a chase race between various scientists in search of the most precise value possible, which continues to this day. The most accredited today, which in physics is indicated by the letter c, is equal to 299792.458 kilometers per second. We observe incidentally that this c is the same that appears in the famous formula $E=mc2$. We will find it several times, because it is one of the main pieces of that great puzzle called universe changed.

Thomas Young

In that year an English doctor with many interests, including physics, performed an experiment that would go down in history. Thomas Young (1773-1829) was a child prodigy: he learned to read at the age of two and by the age of six he had already read the entire Bible twice and had begun studying Latin.3 It was then the turn of Greek, French and Italian. Soon he was confronted with philosophy, natural history and mathematical analysis invented by Newton; he also learned to build microscopes and telescopes. Before the age of twenty, he learned Hebrew, Chaldean, Aramaic, the Samaritan variant of biblical Hebrew, Turkish, Parse and Amharic. From 1792 to 1799 he studied medicine in London, Edinburgh and Göttingen, where, forgetting his Quaker education, he was also interested in music, dance and theater. He boasted that he had never lazy one day. Obsessed by ancient Egypt, this extraordinary gentleman, amateur and self-taught, was among the first to translate hieroglyphics. The compilation of the dictionary of ancient Egyptian languages was a feat that kept him literally busy until the day of his death.

His career as a doctor was much less fortunate, perhaps because it did not instill confidence in the sick or because he lacked the je ne sais quoi that is needed in relations with patients. The lack of attendance at his London clinic, however, allowed him to take time to attend meetings of the Royal Society and to discuss with the major scientific figures of the time. As far as we are interested here, his greatest discoveries were in the field of optics. He began researching the subject in 1800 and within seven years set up an extraordinary series of experiments that seemed to confirm the undulating theory of light with increasing confidence. But before we get to describe the most famous one, we must take a look at the waves and their behavior.

Take for example those of the sea, so loved by surfers and romantic poets. Let's look at them offshore, free to travel. The distance between two consecutive ridges (or between two bellies) is called wavelength, while the height of the ridge in relation to the surface of the calm sea is called amplitude. The waves move at a certain speed, which in the case of light, as we have already seen, is indicated with c. Let's fix it in one point: the period between the passage of one crest and the next is a cycle. The frequency is the speed at which the cycles are repeated; if for example we see three crests pass in one minute, let's say the frequency of that wave is 3 cycles/minute. We have that the wavelength multiplied by the frequency is equal to the speed of the wave itself; for example, if the wave of 3 cycles/minute has a wavelength of 30 meters, this means that it is moving at 90 meters per minute, equal to 5.4 kilometers per hour.

We now see a very familiar type of waves, those sound waves. They appear in various frequencies. Those audible to the human ear range from 30 cycles/second of the lowest sounds to 17000 cycles/second of those above. The central note, or the3, is set at 440 cycles/second. The speed of sound in the air, as we have already seen, is about 1200 km/hour. Thanks to simple calculations and remembering that the wavelength is equal to the speed divided by frequency, we deduce that the wavelength of la3 is (330 meters/second): (440 cycles/second) = 0.75 meters. The wavelengths audible by humans are (330 meters/second) : (440 cycles/second) = 0.75 meters: (17 000 cycles/second) = 2 centimeters to (330 meters/second) : (30 cycles/second) = 11 meters. It is this parameter, together with the speed of sound, that determines what happens to sound waves when they echo in a gorge, or propagate in a large open space such as a stadium, or reach the audience of a theater.

In nature there are many types of waves: in addition to marine and sound waves, we remember for example the vibrations of strings and seismic waves that shake the earth beneath our feet. All of them can be described well with classical (not quantum) physics. The amplitudes refer from time to time to different measurable quantities: the height of the wave above sea level, the intensity of sound waves, the displacement of the rope from the quiet state or the compression of a spring. In any case we are always in the presence of a perturbation, a deviation from the norm within a transmission medium that was previously quiet. The perturbation, which we can visualize as the pinch given to a string, propagates in the form of a wave. In the realm of classical physics the energy transported by this process is determined by the amplitude of the wave.

Sitting in his little boat in the middle of a lake, a fisherman threw the line. A float is visible on the surface, which serves both to prevent the hook from going to the bottom and to signal that something has taken the bait. The water is rippled, and the float goes up and down following the waves. Its position changes regularly: from level zero to a crest, then back to level zero, then back to a belly, then back to level zero and so on. This cyclic motion is given by a wave called harmonic or sinusoidal. Here we will simply call it "wave".

Open problems

The theory, at the time, was not able to answer satisfactorily various questions: what exactly is the mechanism by which light is generated? how does absorption take place and why do colored objects absorb only certain precise wavelengths, i.e. colors? what mysterious operation inside the retina allows us to "see"? All questions that had to do with the interaction between light and matter. In this regard, what is the mode by which light propagates in empty space, as between the Sun and the Earth? The analogy with sound and material waves would lead us to think that there is a medium through which perturbation occurs, a mysterious transparent and weightless substance that pervades deep space. In the nineteenth century it was assumed that this substance really existed and it was called ether.

Then there is still a mystery about our star. This colossal light generator produces both visible and invisible light, where by "invisible light" we mean light with a wavelength too long (from infrared on) or too short (from ultraviolet down) to be observed. The Earth's atmosphere, primarily the ozone layer of the upper stratosphere, blocks much of the ultraviolet rays and even shorter waves, such as X-rays. Now let's imagine that we have invented a device that allows us without too many complications to absorb light selectively, only in certain frequencies, and to measure its energy.

This device exists, yes it does (it is even present in the best equipped laboratories of high schools) and is called a spectrometer. It is the evolution of the Newtonian prism, able to decompose light into various colors by selectively diverting its components according to various angles. If we insert a mechanism that allows a quantitative measurement of these angles, we can also determine the respective wavelengths (which depend directly on the angles themselves).

Let us now concentrate on the point where dark red fades into black, i.e. at the border of visible light. The scale of the spectrometer tells us that we are at 7500 Å, where the letter "Å" is the symbol of the angstrom, a unit of length named in honor of the Swedish physicist Anders Jonas Ångström, one of the pioneers of spectroscopy. An angstrom is equal to 10^{-8} centimeters, or one hundred millionth of a centimeter. We have therefore discovered that between two ridges of light waves at the edge of the visible run 7500 Å, that is 7.5 thousandths of a centimeter. For lengths longer than this we need instruments sensitive to infrared and long waves. If instead we go to the other side of the visible spectrum, on the violet side, we see that the corresponding wavelength is about 3500 Å. Below this value the eyes do not help us and we need to use other instruments.

So far everything is fine, we are just specifying the results obtained by Newton on the decomposition of light. In 1802, however, the English chemist William Wollaston pointed a spectrometer in the direction of sunlight and discovered that in addition to the spectrum of colors neatly arranged from red to violet there were many dark and thin lines. What was this?

At this point Joseph Fraunhofer (1787-1826), a Bavarian with great talent and little formal education, a skilled lens maker and optics expert, entered the scene.10 Eleventh and last son of a glassmaker who fell into poverty, Fraunhofer barely finished elementary school and was then sent to work in his family's workshop in Dickensian conditions. After the death of his father, the sickly boy found unqualified employment as an apprentice at a glassmaker and mirror manufacturer in Munich. In 1806 he managed to join an optical instrument company in the same city, where thanks to the help of an astronomer and a skilled craftsman he learned the secrets of optics to perfection and developed a mathematical culture. Frustrated by the poor quality of the glass at his disposal, the perfectionist Fraunhofer tore up a contract that allowed him to peek into the jealously guarded industrial secrets of a famous Swiss glassworks that had recently moved its activities to Munich. This collaboration gave rise to technically advanced lenses and, above all, for what we are interested in here, a fundamental discovery that would ensure Fraunhofer a place in the history of science.

In his search for the perfect lens, he came up with the idea of using the spectrometer to measure the refractive capacity of various types of glass. As he examined the decomposition of sunlight, he realized that the black lines discovered by Wollaston were really many, about six hundred. He began to catalogue them systematically by wavelength, and by 1815 he had examined almost all of them. The most evident were labeled with capital letters from A to I, where A was a black line in the red area and I was at the extreme limit of the violet. What were they caused by? Fraunhofer was aware of the phenomenon whereby certain metals or salts, when exposed to fire, emit light of precise colors; he measured these rays with the spectrometer and saw many clear lines appear in the region of the wavelengths corresponding to the color emitted.

The interesting thing was that their structure was identical to that of the black lines in the solar spectrum. The table salt, for example, had many light lines in the region that Fraunhofer had marked with the letter D. For an explanatory model of the phenomenon we had to wait a little longer. As we know, each well defined wavelength corresponds to a uniquely defined frequency. It had to be a mechanism that made the matter vibrate, presumably at atomic level, according to certain established frequencies. The atoms (which at Fraunhofer's time had not yet been proven to exist) left macroscopic imprints.

The imprints of atoms

As we have seen above, a tuning fork set to "give the go" vibrates at a frequency of 440 cycles per second. In the microscopic realm of atoms the frequencies are immensely higher, but already in Fraunhofer's time it was possible to imagine a mechanism by which the mysterious particles were equipped with many very small equivalents of the tuning fork, each with its own characteristic frequency and able to vibrate and emit light with a wavelength corresponding to the frequency itself.

And why then appear the black lines? If the sodium atoms excited by the heat of the flame vibrate with frequencies that emit light between 5911 and 5962 Å (values that correspond to shades of yellow), it is likely that, in reverse, they prefer to absorb light with the same wavelengths. The hot surface of the Sun emits light of all kinds, but it then passes through the "corona", i.e. the less hot gases of the solar atmosphere. This is where selective absorption occurs by the atoms, each of which retains light of the wavelength congenial to it; this mechanism is responsible for the strange black lines observed by Fraunhofer. One piece at a time, subsequent research has revealed that each element, when excited by heat, emits a characteristic series of spectral lines, some sharp and sharp (such as bright red neon), others weak (such as the blue of mercury vapor lamps). These lines are the fingerprints of the elements, and their discovery was a first indication of the existence of mechanisms similar to the "tuning forks" seen above (or some other devilry) within atoms.

The spectral lines are very well defined, so it is possible to calibrate the spectrometer in order to provide very precise results, distinguishing for example a light with a wavelength of 6503.2 Å (dark red) from another to 6122.7 Å (light red). At the end of the 19th century, thick tomes were published listing the spectra of all the elements then known, thanks to which the most skilled spectroscopy experts were able to determine the chemical composition of unknown compounds and to recognize even the smallest contamination. Yet no one had any idea what was the mechanism responsible for producing such clear messages. How the atom worked remained a mystery.

Another success of spectroscopy was of a deeper nature. In the Sun fingerprint we could read, incredibly, many elements present on Earth: hydrogen, helium, lithium and so on. When we started to analyze the light coming from stars and distant galaxies, we realized that the result was similar. The universe is composed in every place by the same elements, which follow the same laws of nature, which suggests that everything had a unique origin thanks to a mysterious physical process of creation.

In parallel, between the seventeenth and nineteenth centuries science was trying to solve another problem: how do forces, and particularly gravity, transmit their action over great distances? If we attach a carriage to a horse, we see that the force used by the animal to pull the vehicle is transmitted directly, through the harnesses and rods. But how does the Earth "feel" the Sun, which is 150 million kilometers away? How does a magnet attract a nail placed at a certain distance? In these cases there are no visible connections, so we must assume a mysterious "action at a distance". According to Newton's formulation, gravity acts at a distance, but it is not given to know what is the "rod" that connects two bodies like the Earth and the Sun. After having vainly struggled with this problem, even the great English physicist had to surrender and let posterity take care of the matter.

What is a black body and why do we care so much?

All bodies emit energy and absorb it from their surroundings. Here for "body" we mean a great object, that is macroscopic, composed from many billions of atoms. The higher its temperature, the greater the amount of energy emitted.

The warm bodies, in all their parts (which we can consider in turn as bodies), tend to reach a balance between the value of the energy given to the external environment and that absorbed. If, for example, you take an egg from the fridge and plunge it into a pot full of boiling water, the egg gets hot and the temperature of the water decreases. Conversely, if you throw a hot egg in cold water the heat transfer takes place in the opposite direction. If you do not provide more energy, after a while egg and water will be at the same temperature. It is a home experiment easy to perform, which clearly illustrates the behavior of the bodies with respect to heat. The final state in which egg and water temperatures are equal is called thermal equilibrium, and is a universal phenomenon: a hot object immersed in a cold environment cools down, and vice versa. At thermal equilibrium, all the parts of which the body is made up are at the same temperature, therefore they emit and absorb energy in the same way.

When you are lying on a beach on a beautiful day, your body is emitting and absorbing electromagnetic radiation: on the one hand you absorb energy produced by the primeval radiator, the Sun, and on the other hand you emit a certain amount of heat because the body has regulation mechanisms that allow it to maintain the right internal temperature.1 A healthy adult man, at a temperature of 37 °C, radiates about 100 watts into the surrounding environment. The various parts of the body, from the liver to the brain, from the heart to the fingertips, are kept in thermal equilibrium so that biochemical processes take place without any problems. If the environment is very cold, the body must produce more energy, or at least not disperse it around, if it wants to maintain the ideal temperature. The blood flow, which is responsible for the transfer of heat to the surface of the body, is therefore reduced so that internal organs do not lose heat, which is why we feel cold in our fingers and nose. On the contrary, when the environment is very hot, the body has to increase the dispersed energy, which happens thanks to sweat: the evaporation of this hot liquid on the skin involves the use of an additional amount of energy of the body (a sort of air conditioning effect), which is then dispersed outside. The fact that the human body radiates heat is evident in closed and crowded rooms: thirty people piled in a meeting room produce 3 kilowatts, which are able to heat the environment quickly. On the contrary, in the Antarctic those same boring colleagues could save your life if you embrace them tightly, as emperor penguins do to protect their fragile eggs from the rigours of the long winter.

Humans, penguins and even toasters are complex systems that produce energy from within. In our case, the fuel is given by food or fat stored in the body; a toaster instead has as its energy source the collisions of electrons in the electric current with the heavy metal atoms of which the resistance is made. The electromagnetic radiation emitted, in both cases, is dispersed in the external environment through the surface in contact with air, in our case the skin. This radiation usually has a color that is the imprint of particular "atomic transitions", that is, it is the daughter of chemistry. Fireworks, for example, when they explode are certainly hot and the light they emit depends on the nature of the compounds they contain (strontium chloride, barium chloride and others),2 thanks to whose oxidation they shine bright and spectacular colors.

These particular cases are fascinating, but the electromagnetic radiation always behaves in the same way, in any system, in the simplest case: the one in which all the chromatic effects due to the various atoms are mixed and cancelled, giving life to what physicists call thermal radiation. The ideal object that produces it is called black body. It is therefore a body that by definition produces only thermal radiation when heated, without a particular color prevailing - and without the special effects of fireworks. Even if it is an abstract concept, there are objects of daily use that can be approximated rather well to an ideal black body. For example, the Sun emits light with a well-defined spectrum (the Fraunhofer lines), due to the presence of various types of atoms in the surrounding gaseous corona; but if we consider the radiation as a whole we see that it is very similar to that of a (very hot) black body. The same can be said for the hot embers, the toaster resistances, the Earth's atmosphere, the mushroom of a nuclear explosion and the primordial universe: all reasonable approximations of a black body.

A very good model is given by an old style coal boiler, like for example the one present in steam trains, that as the temperature rises produces inside it practically pure thermal radiation. In fact it was this the model used from the physicists of end XIX century for the study of the black body. To have a source of pure thermal radiation, it must be isolated in some way from the heat source, in this case the burning coal. For this purpose we build a robust and thick-walled container, say iron, in which we drill a hole to observe what happens inside and make measurements. Let's put it in the boiler, let it heat up and peek through the hole. We detect a pure thermal radiation, which fills the entire container. This is emitted from the hot walls and bounces from one end to the other; a small part comes out from the observation hole.

With the help of a few instruments we can study the heat radiation and check to what extent the various colors (i.e. the various wavelengths) are present. We can also measure how the composition changes when the temperature of the boiler changes, i.e. study the radiation in thermal equilibrium.

At the beginning the hole emits only lukewarm and invisible infrared radiation. When we turn up the heating we see a dark red light that looks like the one visible inside the toaster. Increasing the temperature again the red becomes brighter, until it becomes yellow. With a particularly powerful machine, like the Bessemer converter of steel mills (where oxygen is introduced), we can reach very high temperatures and see how the radiation becomes practically white. If we could use an even stronger heat source (therefore not a classic boiler, which would melt), we would observe a bright light tending to blue coming out from the hole at very high temperature. We have reached the level of nuclear explosions or bright stars like Rigel, the blue supergiant of the Orion constellation which is the highest energy source of thermal radiation in our galactic vicinity.

The study of thermal radiation was an important field of research, at the time completely new, which combined two different subjects: the study of heat and thermal balance, or thermodynamics, and electromagnetic radiation. The collected data seemed completely harmless and gave anyway the possibility to make interesting researches. Nobody could suspect that they were important clues in what would soon become the scientific yellow of the millennium: the quantum properties of light and atoms (that in the end are those that do all the work).

Chapter 4
Mr. Planck

The classical theory of light and Planck's calculations led not only to the conclusion that the distribution of wavelengths was concentrated in the blue-violet parts, but even (due to the desperation of theoretical physicists, who were increasingly perplexed) that the intensity became infinite in the more remote regions of ultraviolet. There was someone, perhaps a journalist, who called the situation an "ultraviolet catastrophe". In fact it was a disaster, because the theoretical prediction did not agree at all with the experimental data. To listen to the calculations, the embers would not emit red light, as humanity has known for at least a hundred thousand years, but blue light.

It was one of the first cracks in the building of classical physics, which until then seemed unassailable. (Gibbs had found another one, probably the first ever, about twenty-five years earlier; at the time its importance had not been understood, except perhaps by Maxwell). The black body radiation curves shown in figure 4.1 have peaks that depend on temperature (more towards red at the low ones, more towards blue at the high ones). All of them, however, go down quickly to zero in the very short wave area. What happens when an elegant and well tested theory, conceived by the greatest minds of the time and certified by all European academies, clashes with the brute and crude experimental data? If for religions dogmas are untouchable, for science defective theories are bound to be swept away sooner or later.

Classical physics predicts that the toaster will shine blue when everyone knows it is red. Remember this: every time you make toast, you are observing a phenomenon that blatantly violates classical laws. And even if maybe you do not know it (for now), you are having the experimental confirmation that light is made of discrete particles, it is quantized. It is quantum mechanics live! But, you will object, haven't we just seen in the previous chapter that thanks to the genius of Mr. Young it has been proved that light is a wave? Sure, and it is true. Let's get ready, because things are about to get very strange. We are still travelers exploring new and bizarre worlds far away - and yet we still get there even from a toaster.

Max Planck

Berlin, the epicenter of the ultraviolet catastrophe, was the realm of Max Planck, a theoretical physicist then in his forties, a great thermodynamic expert.7 Fully aware of the catastrophe, he was the first to want to understand something about it. In 1900, starting from the experimental data collected by his colleagues and using a mathematical trick, he managed to transform the formula derived from classical theory into another one that matched the measurements very well. Planck's manipulation allowed the long waves to show themselves quietly at all temperatures, more or less as expected by classical physics, but he cut the short waves imposing a sort of "toll" on their emission. This obstacle limited the presence of blue light, which in fact radiated less abundantly.

The trick seemed to work. The "toll" made the higher frequencies (remember: short waves = high frequencies) were more "expensive", that is they required much more energy than the low ones. So, according to Planck's right reasoning, at low temperatures the energy was not enough to "pay the toll" and short waves were not emitted. To return to our theatrical metaphor, a way had been found to free the front rows and push the spectators towards the middle rows and the tunnels. A sudden intuition (which was not typical of his way of working) allowed Planck to connect wavelength (or frequency equivalent) to energy: the longer the length, the less energy.

It seems an elementary idea, and indeed it is, because that's how nature works. But classical physics did not contemplate it at all. The energy of an electromagnetic wave according to Maxwell's theory depended only on its intensity, not on color or frequency. How did Planck put this characteristic in his treatment of the black body? How did he manage to pass the idea that energy depends on intensity but also on frequency? There is still one piece missing from the puzzle, because you have to specify what has more energy as the frequencies increase.

To solve the problem, Planck found an efficient way to divide the emitted light, whatever the wavelength, into packets called quantum, each of which has a quantity of energy related to its frequency. Planck's illuminating formula is really as simple as possible:

$$E=hf$$

Put in words: "the energy of a quantum of light is directly proportional to its frequency". So the electromagnetic radiation is composed of many small packages, each of which has a certain energy, equal to its frequency multiplied by a constant h. The intensity of the emitted light is equal to the number of those who register at a certain frequency multiplied by their energy. Planck's effort to reconcile the data with the theory led to the idea that high frequencies (i.e. short waves) were expensive in terms of energy for the black body. His equation, at all temperatures, was in perfect harmony with the curves obtained from experimental measurements.

It is interesting to note that Planck did not immediately realize that his modification to Maxwell's theory had directly to do with the nature of light. Instead, he was convinced that the key to the phenomenon was in the atoms that made up the walls of the black body, that is the way the light was emitted. The preference for red over blue was not due, for him, to intrinsic properties of these wavelengths, but to the way the atoms moved and emitted radiation of various colors. In this way he hoped to avoid conflicts with classical theory, which had worked wonders until then: after all, electric motors were pushing trains and streetcars all over Europe and Marconi had just patented the wireless telegraph. Maxwell's theory was obviously not wrong and Planck had no intention of correcting it: better to try to amend the most mysterious thermodynamics.

Yet his hypothesis about thermal radiation involved two sensational deviations from classical physics. First, the correlation between the intensity (i.e. the energy content) of the radiation and its frequency, completely absent in the Maxwellian picture. Then, the introduction of discrete quantities, quanta. These are two aspects related to each other. For Maxwell the intensity was a continuous quantity, able to assume any real value, dependent only on the values of electric and magnetic fields associated with the light wave. For Planck the intensity at a given frequency is equal to the number of quanta corresponding to the frequency itself, each of which carries an energy equal to $E=hf$. It was an idea that smelled suspiciously of "light particles", yet all the diffraction and interference experiments continued to confirm the wave-like nature.

Nobody then, including Planck, fully understood the meaning of this turning point. For their discoverer, the quanta were concentrated impulses of radiation, coming from the atoms of the black body in frantic movement due to thermal agitation, which emitted them according to unknown mechanisms. He could not have known that that h, now called Planck's constant, would become the spark of a revolution that would lead to the first roars of quantum mechanics and modern physics. For the great discovery of the "quantum energy", occurred when he was forty-two years old, Planck was awarded the Nobel Prize for physics in 1918.

Einstein enters the scene

The extraordinary consequences of the introduction of quantum physics were understood immediately afterwards by a young physicist then unknown, none other than Albert Einstein. He read Planck's article in 1900 and, as he later declared, he felt "the earth beneath his feet is missing".8 The underlying problem was this: were the energy packets children of the emission mechanism or were they an intrinsic characteristic of light? Einstein realized that the new theory was deploying a well-defined, disturbingly discreet, particle-like entity that intervened in the process of light emission by superheated substances. At first, however, the young physicist refrained from embracing the idea that quantization was a fundamental characteristic of light.

Here it is necessary to say a few words about Einstein. He was not a child prodigy and did not particularly like school. As a boy, no one would have predicted a successful future for him. But science had always fascinated him, ever since the day his father showed him a compass when he was four years old. He was bewitched by it: invisible forces forced the needle to always point north, in whatever direction it was turned. As he wrote in his old age: "I remember well, or rather I think I remember well, the deep and lasting impression left by this experience. Still at a young age, Einstein was also captivated by the magic of algebra, which he had learned from an uncle, and bewitched by a geometry book read at the age of twelve. At sixteen he wrote his first scientific article, dedicated to the ether in the magnetic field.

At the point where our story has arrived, Einstein is still a stranger. Not having obtained a university assignment of any kind after the end of his studies, he began to give private lectures and make substitutes, and then came to the position of employee at the Swiss Patent Office in Bern. Although he only had weekends free for his research, in the seven years he spent in that office he laid the foundations of 20th century physics and discovered a way to count atoms (i.e. to measure Avogadro's constant), invented narrow relativity (with all its profound consequences on our notions of space and time, without forgetting E=mc2), made important contributions to quantum theory and more. Among his many talents, Einstein could include synesthesia, i.e. the ability to combine data from different senses, for example vision and hearing. When he meditated on a problem, his mental processes were always accompanied by images and he realized that he was on the right track because he felt a tingling sensation at his fingertips. His name would become synonymous with great scientist in 1919, when thanks to a solar eclipse there was experimental confirmation of his theory of general relativity. The Nobel Prize, however, obtained him for a work of 1905, different from relativity: the explanation of the photoelectric effect.

Just imagine the culture shock experienced by physicists in 1900, who were quiet and serene in their studies to consult the data on the continuous spectra of radiation emitted by hot objects, data that had accumulated for almost half a century. The experiments that had produced them were possible thanks to Maxwell's theory of electromagnetism, accepted by about thirty years, which predicted that light was a wave. The fact that a phenomenon so typically wave-like could, in some circumstances, behave as if it was composed of discrete packets of energy, in other words "particles", threw the scientific community in a terrible state of confusion. Planck and his colleagues, however, assumed that sooner or later they would arrive at a high level, so to speak, neoclassical explanation. After all, black body radiation was a very complicated phenomenon, like atmospheric weather, in which many events in themselves simple to describe come together in an apparently elusive complex state. But perhaps the most incomprehensible aspect of it was the way nature seemed to reveal for the first time, to those who had the patience to observe it, its most intimate secrets.

Arthur Compton

In 1923 the particle hypothesis marked a point in its favor thanks to the work of Arthur Compton, who began to study the photoelectric effect with X-rays (very short-wave light). The results obtained by him did not lie: the photons that collided with electrons behaved like particles, that is like small billiard balls.[11] The electron at first stationary splashes off and the photon that is thrown against it rebounds. This phenomenon is now called "Compton effect" or more properly "scattering Compton".

As in all elastic collisions in classical physics, during this process the total energy and the momentum of the electron + photon system are preserved. But in order to fully understand what happens, it is necessary to break the delay and treat the photon as a particle to all effects, step to which Compton arrived gradually, after having noticed the failure of all his previous assumptions. In 1923 the nascent quantum theory (Niels Bohr's "old quantum theory") was not yet able to explain the Compton effect, which would be understood only thanks to subsequent developments. When the American physicist presented his results at a conference of the American Physical Society, he had to face the open opposition of many colleagues.

As a good son of a family from the Mennonite minority in Wooster, Ohio, used to work hard, Compton did not lose heart and continued to refine his experiments and interpretations of the results. The final confrontation took place in 1924, during a seminar of the British Association for the Advancement of Science specifically organized in Toronto. Compton was very convincing. His arch-enemy, William Duane of Harvard, who until then had not been able to replicate his results, returned to the laboratory and redo the controversial experiment himself. In the end he was forced to admit that the Compton effect was true. His discoverer won the Nobel Prize in 1927. Compton was one of the main architects of the development of American physics in the twentieth century, so much so that he was given the honor of a cover of the "Times" on January 13, 1936.

Summing up, what do these results show? On the one hand, we are faced with various phenomena that show that light seems to consist of a stream of particles, how many luminous called photons (poor Newton, if only he had known...). On the other hand we have Young's experiment with the double slit (and millions of other experiments that confirm it, still performed in school laboratories all over the world), thanks to which light shows to behave like a wave. Three hundred years after the wave-particle dispute, we are still back to the beginning. Is it perhaps an irresolvable paradox? How can an entity be simultaneously a wave and a particle? Do we have to give up physics and dedicate ourselves to Zen and motorcycle maintenance?

Glass and mirrors

The photon, as a particle, simply "is there". It triggers detectors, collides with other particles, explains the photoelectric effect and the Compton effect. It exists! But it does not explain the interference - and also another phenomenon.

You will remember that in chapter 1 we stopped in front of a store window full of sexy underwear. Now we continue our walk and we arrive at the windows of a department store, where the spring-summer collection is displayed on elegant mannequins. The sun is shining and the contents of the window is clearly visible; but on the glass we realize that there is also a faint reflection of the street and passers-by, including us. By chance, in this showcase is also contained a mirror, which reflects in detail our image. So we see ourselves twice: clearly in the mirror and weakly in the glass.

Here is a plausible explanation: the sun's rays reflect on the surface of our body, pass through the showcase, hit the mirror and come back until they reach our retina. A small percentage of the light, however, comes also reflected from the same showcase. Well, so what? All this is perfectly logical, however it is seen. If the light is a wave, no problem: the waves are normally subject to reflection and partial refraction. If the light, instead, is constituted from a flow of particles we can explain the all admitting that a certain part of the photons, we say the 96 per cent, passes through the glass and the remaining 4 per cent comes reflected. But if we take a single photon of this huge stream, consisting of particles all the same, how do we know how it will behave in front of the glass? How does our photon (let's call it Bernie) decide which way to go?

Now let's imagine this horde of identical particles heading towards the window. The great majority crosses it, but some from time to time is rejected. We remember that the photons give themselves for indivisible and irreducible - nobody has ever seen in nature 96 per cent of a photon. So Bernie has two alternatives:

either he passes through in one piece, or he is rejected in one piece. In the latter case, which occurs 4% of the time, perhaps he collided with one of the many glass atoms. But if the things were so we would not see our image reflected in the weak but well delineated showcase; we would observe instead a glass slightly tarnished from that 4 per cent of lost photons. The image, that we recognize without problems as "ours", indicates that we are in presence of a coherent and undulating phenomenon, and yet the photons exist. Here we are faced with another problem, that of partial reflection. It seems that there is a 4% probability that a photon, understood as a particle, ends up in a wave that is reflected. That Planck's hypotheses led to the introduction of random and probabilistic elements in physics was already clear to Einstein in 1901. He did not like it at all, and over time his disgust would grow.

The walrus and the panettone

As if Planck's solution to the ultraviolet catastrophe problem and Einstein's explanation of the photoelectric effect were not shocking enough, classical physics was faced with a third wake-up call in the early 20th century: the failure of Thomson's atomic model, or "panettone model".

Ernest Rutherford (1871-1937) was a ruffled big man who looked like a walrus. After winning the Nobel Prize in Chemistry for his research in radioactivity, in 1917 he became director of the prestigious Cavendish Laboratory in Cambridge. He was born in New Zealand into a large farming family; life on the farm had accustomed him to hard work and made him a resourceful man. A lover of machines and technological innovation, he had dabbled since childhood in fixing watches and building working models of water mills. In his postgraduate studies, he was involved in electromagnetism and was able to build a radio wave detector before Marconi conducted his famous experiments. Thanks to a scholarship he arrived at Cambridge, where his radio, capable of picking up signals from almost a kilometer away, made a favorable impression on many professors, including J. J. Thomson, who at the time ran the Cavendish Laboratory.

Thomson invited Rutherford to work with him on one of the novelties of the time, X-rays, then known as Becquerel rays, and to study the phenomenon of electrical discharge in gases. The young kiwi was homesick, but that was a must-have offer. The result of their collaboration was summarized in a famous article on ionization, which was explained by the fact that X-rays, colliding with matter, seemed to create an equal number of electrically charged particles, called "ions". Thomson would then publicly claim that he never knew anyone as skilled and passionate about research as his student.

Around 1909 Rutherford coordinated a working group dedicated to the so-called alpha particles, which were fired against a thin sheet of gold to see how their trajectories were diverted by the heavy metal atoms. In those experiments something unexpected happened. Almost all the particles were diverted a little bit, while passing through the gold foil, directed to a detection screen at a certain distance. But one in every eight thousand bounced and did not go beyond the foil. As Rutherford then said, "it was like firing a mortar at a sheet of paper and seeing the bullet come back. What was happening? Was there something inside the metal that could repel alpha, heavy, positively charged particles?

Thanks to previous research by J.J. Thomson, at that time it was known that atoms contained electrons, light and electrically negative. To make the construction stable and to balance it all, of course, it was necessary the presence of an equal and opposite amount of positive charges. Where they were, however, was then a mystery. Before Rutherford, no one had been able to give a shape inside the atom.

In 1905 J. J. Thomson had proposed a model that provided a positive charge spread in a homogeneous way inside the atom and the various electrons scattered around like raisins in the panettone - for this reason it was baptized by the physicists "panettone model" (plum pudding model in English). If the atom was really made like this, the alpha particles of the above experiment should always pass through the foil: it would be like shooting bullets on a shaving foam veil. Here, now imagine that in this situation one bullet out of every eight thousand is deflected by the foam until it comes back. This happened in Rutherford's laboratory.

According to his calculations, the only way to explain the phenomenon was to admit that all the mass and the positive charge of the atom were concentrated in a "nucleus", a small ball located in the center of the atom itself. In this way there would have been the mass and charge concentrations necessary to repel the alpha particles, heavy and positive, that eventually arrived on a collision course. It was as if inside the shaving foam veil there were many hard and resistant balls, able to deflect and repel the bullets. The electrons in turn were not scattered but they went around the nucleus. Thanks to Rutherford, therefore, the panettone model was delivered to the dustbin of history. The atom looked rather like a small solar system, with miniature planets (the electrons) orbiting around a dense and dark star (the nucleus), all held together by electromagnetic forces.

With further experiments it was discovered that the nucleus was really tiny: its volume was about one thousandth of billionth of the atom. For contrast, it contained in itself more than 99,98 percent of the total mass of the atom. So the matter was largely made of vacuum, of dots around which electrons were whirling around and at great distance. Really incredible: matter is basically made of nothing! (also the "solid" armchair you are sitting on now is almost all empty). At the time of this discovery the classical physics, from Newton's F=ma to Maxwell's laws, was still considered unassailable, both at microscopic level and at large scale, at solar system level. It was believed that in the atom were at work the same laws valid elsewhere. Everyone slept peaceful dreams, until Niels Bohr arrived.

The melancholy Dane

One day, Niels Bohr, a young Danish theorist who was perfecting at the time at the Cavendish Laboratory, attended a conference in Rutherford and was so impressed by the great experimenter's new atomic theory that he asked him to work with him at the University of Manchester, where he was then. He agreed to host him for four months in 1912.

Reflecting calmly on the new experimental data, Bohr soon realized that there was something wrong with the model. In fact, it was a disaster! If you apply Maxwell equations to an electron in very fast circular orbit around the nucleus, it results that the particle loses almost immediately all its energy, in the form of electromagnetic waves. Because of this the orbital ray becomes smaller and smaller, reducing to zero in only 10-16 seconds (one ten millionth of a billionth of a second). In few words, an electron according to classical physics should fall almost instantly on the nucleus. So the atom, that is matter, is unstable, and the world as we know it is physically impossible. Maxwell equations seemed to involve the collapse of the orbital model. So the model was wrong, or were the venerable laws of classical physics.

Bohr started to study the simplest atom of all, the hydrogen one, that in Rutherford's model is a single negative electron orbiting around the positive nucleus. Thinking back to Planck and Einstein's results, and to some ideas that were in the air about the wave behavior of particles, the young Dane launched in a very little classical and very risky hypothesis. According to Bohr electron are allowed only certain orbits, because their motion inside the atom is similar to that of waves. Among the allowed orbits there is one of minimum energy level, where the electron is as close as possible to the nucleus: the particle cannot go down more than that and therefore it cannot emit energy while jumping to a lower level - which does not exist at all. This special configuration is called fundamental state.

With his model, Bohr tried first of all to explain theoretically the discrete spectrum of atoms, those more or less dark lines that we have already met. As you will remember, the various elements, if heated until they emit light, leave on the spectrometer a characteristic imprint consisting of a series of colored lines that stand out clearly against a darker background. In the spectrum of sunlight, then, there are also black and thin lines at certain precise points. The light lines correspond to the emissions, the dark lines to the absorptions. Hydrogen, like all elements, has its spectral "fingerprint": to these data, known at the time, Bohr was trying to apply his newborn model.

In three subsequent articles published in 1913, the Danish physicist enunciated his daring quantum theory of hydrogen atom. The orbits allowed to the electron are characterized by fixed amounts of energy, that we will call E_1, E_2, E_3 etc. An electron emits radiation when it "jumps" from a higher level, let's say E_3, to a lower one, let's say E_2: it is a photon whose energy (given, remember, from $E=hf$) is equal to the difference between the two levels. So $E_2-E_3=hf$. Adding this effect in the billions of atoms where the process happens at the same time, we obtain as result the clear lines of the spectrum. Thanks to a model that kept in part the Newtonian mechanics but it differed when it disagreed with the experimental data, Bohr was triumphantly able to calculate the wavelengths corresponding to all spectral lines of hydrogen. His formulas depended only on known constants and values, such as mass and electron charge (as usual seasoned here and there by symbols like π and, of course, the hallmark of quantum mechanics, Planck constant h).

Summing up, in Bohr's model the electron remains confined in few allowed orbits, as if by magic, that correspond to well defined energy levels E_1, E_2, E_3 etc. The electron can absorb energy only in "packets" or "quanta"; if it absorbs a sufficient number, it can "jump" from the level where it is at to a higher one, for example from E_2 to E_3; vice versa, electrons in the higher levels can spontaneously slide down, returning for example from E_3 to E_2, and in doing so they emit quanta of light, that is photons. These photons can be observed because they have specific wavelengths, which correspond to the spectral lines. Their values are predicted exactly, in the case of hydrogen atom, by Bohr model.

The character of the atom

So it is thanks to Rutherford and Bohr if today the most known representation of the atom is the solar system one, where small electrons whizz around the nucleus like many small planets, on orbits similar to the elliptical ones foreseen by Kepler. Many maybe think that the model is precise and the atom is made just like that. Alas no, because Bohr's intuitions were brilliant but not entirely correct. The proclamation of the triumph was premature. It was realized that his model was applied only to the simplest atom, the hydrogen one, but it failed already at the next step, with helium, atom equipped with two electrons. The twenties were just around the corner and quantum mechanics seemed stuck. Only the first step had been done, corresponding to what today we call "old quantum theory".

The founding fathers, Planck, Einstein, Rutherford and Bohr, had started the revolution but had not yet reaped the fruits. It was clear to everyone that innocence was lost and that physics was becoming strange and mysterious: there was a world of packet energy and electrons magically jumping only on certain orbits and not on others, a world where photons are waves and particles at the same time, without being at the bottom neither one nor the other. There was still a lot to understand.

Chapter 5
An uncertain Heisenberg

A nd here is the moment you've all been waiting for. We are about to face the real quantum mechanics head-on and enter into alien and disconcerting territory. The new science pushed to exasperation also Wolfgang Pauli, one of the greatest physicist ever, who in 1925 in a letter to a colleague said he was ready to give up the fight: "Physics is now too difficult. I would rather be a comic actor, or something like that, than a physicist". If such a giant of scientific thought had abandoned research to become the Jerry Lewis of his time, today we would not be talking about the "Pauli exclusion principle" and the history of science could have taken a very different turn.1 Fortunately Pauli held out, as we hope you will do in the next pages. The journey we are about to embark on is not recommended for the faint of heart, but reaching the destination will be an extraordinary reward.

Nature is made in packages

Let's start again from the old quantum theory, formulated by Bohr to account for the results of the Rutherford experiment. As you will remember, it replaced the "panettone" atom model with the idea that there was a dense central nucleus surrounded by whizzing electrons - a configuration similar to the solar system, with our star at the center and the planets orbiting around it. We have already said that also this model has passed to better life. Victim of subsequent refinements, the old quantum theory, with its crazy mixture of classical mechanics and ad hoc quantum adjustments, was at some point completely abandoned. The merit of Bohr was however to present to the world for the first time a quantum atom model, which gained credibility thanks to the results of the brilliant experiment that we will see shortly.

According to the classical laws, no electron could ever remain in orbit around the nucleus. Its motion would be accelerated, like all circular motions (because the speed continuously changes direction in time), and according to Maxwell's laws every charged particle in accelerated motion emits energy in the form of electromagnetic radiation, that is light. According to classical calculations, an electron in orbit would lose almost immediately its energy, that would go away in the form of electromagnetic radiation; therefore the particle in question would lose altitude and would soon crash against the nucleus. The classic atom could not exist, if not in collapsed form, therefore chemically dead and unserviceable. The classical theory was not able to justify the energetic values of electrons and nuclei. It was therefore necessary to invent a new model: the quantum theory.

Moreover, as it was already known at the end of 1800 thanks to the spectral lines data, atoms emit light but only with defined colors, that is with wavelengths (or frequencies) at discrete, quantized values. It seems almost that only particular orbits can exist, and electrons are bound to jump from one to another every time they emit or absorb energy. If it was true the "Keplerian" model of the atom as solar system, the spectrum of emitted radiation would be continuous, because the classical mechanics allows the existence of a continuous range of orbits. It seems instead that the atomic world is "discrete", very different from the continuity provided by Newtonian physics.

Bohr focused his attention on the simplest atom of all, the hydrogen one, with only one proton in the nucleus and one electron orbiting around it. Playing a bit with the new ideas of quantum mechanics, he realized that he could apply to electrons the Planck hypothesis, that is to associate to a certain wavelength (or frequency) the momentum (or energy) of a photon, from which it could be deduced the existence of discrete orbits. After several attempts, finally came to the right formula. Bohr "special" orbits were circular and each one had an assigned circumference, always equal to the quantum wavelength of the electron derived from Planck equation. These magic orbits corresponded to particular energy values, therefore the atom could have only a discrete series of energy states.

Bohr immediately understood that there was a minimum orbit, along which the electron was as close as possible to the nucleus.

From this level it could not fall any lower, therefore the atom was not going to collapse and its fatal destiny. This minimum orbit is known as fundamental state and it corresponds to the minimum energy state of the electron. Its existence implies the stability of the atom. Today we know that this property characterizes all quantum systems.

Bohr's hypothesis was really effective: from the new equations all the numbers that corresponded to the values observed in the experiments were coming out one after the other. Atomic electrons are, as physicists say, "bound" and without the energy supply from the outside they continue to turn around the nucleus. The amount of energy needed to make them to blow away and free them from the atomic bond is called bonding energy, and it depends on the orbit in which the particle is located. (Usually we mean this energy as the minimum required to move an electron away from the atom and bring it to infinite distance and with zero kinetic energy, which we conventionally say is zero energy; but it is just a convention). Vice versa, if a free electron is captured by an atom, it releases a quantity of energy, in the form of photons, equal to the bonding of the orbit in which it ends up.

The bonding energies of orbits (i.e. states) are usually measured in units called electronvolt (symbol: eV). The fundamental state in hydrogen atom, that corresponds to that special orbit of minimum distance from nucleus and maximum bonding energy, has energy equal to 13,6 eV. This value can be obtained theoretically also thanks to the so-called "Rydberg's formula", so called in honor of the Swedish physicist Johannes Rydberg, who in 1888 (extending some researches by Johann Balmer and others) had advanced an empirical explanation of the spectral lines of hydrogen and other atoms. In fact the value of 13,6 eV and the formula from which it can be derived were known for some years, but Bohr was the first to give a rigorous theoretical justification.

The quantum states of an electron in hydrogen atom (equivalent to one of Bohr's orbits) are represented with an integer number n = 1, 2, 3, ... The state with the highest bonding energy, the fundamental one, corresponds to n=1; the first excited state to n=2, and so on. The fact that this discrete set of states is the only one possible in atoms is the essence of quantum theory. The number n has the honor of a name all its own in physics and is called "main quantum number". Each state, or quantum number, is characterized by an energetic value (in eV, like the one seen above) and is labeled with the letters E_1, E_2, E_3 etc.. (see note 3).

As you will remember, in this theory, outdated but not forgotten, it is expected that electrons emit photons jumping from a higher energy state to a lower energy state. This rule obviously does not apply to the fundamental state E_1, that is when n=1, because in this case the electron has no lower orbit available. These transitions happen in a completely predictable and logical way. If for example the electron in the state n=3 goes down to the state n=2, the occupant of this last orbit must level down to n=1. Each jump is accompanied by the emission of a photon with energy equal to the difference between the energies of the involved states, as for example E2-E3 or E1-E2. In case of hydrogen atom, the corresponding numerical values are 10,5 eV - 9,2 eV = 1,3 eV, and 13,6 eV - 10,5 eV = 3,1 eV. Since energy E and wavelength λ of a photon are related by Planck's formula $E=hf=hc/\lambda$, it is possible to derive the energy of emitted photons by measuring their wavelength thanks to spectroscopy. At Bohr's time the accounts seemed to come back for what concerns the hydrogen atom, the simplest one (only one electron around a proton), but already in front of helium, the second element in order of simplicity, we did not know well how to proceed.

In Bohr came another idea, that is to measure the electron momentum through the energy absorption by the atoms, reversing the reasoning seen above. If the hypothesis of quantum states is true, then atoms can acquire energy only in packets corresponding to the differences between states energies, E2 - E3, E1 - E2 and so on. The crucial experiment for the verification of this hypothesis was conducted in 1914 by James Franck and Gustav Hertz in Berlin, and was perhaps the last important research conducted in Germany before the outbreak of the First World War. The two scientists obtained results perfectly compatible with Bohr's theory, of which they were unaware. They would have known the results of the great Danish physicist only several years later.

The terrible twenties

It is difficult to realize the panic that spread among the world's greatest physicists at the beginning of the terrible 1920s, between 1920 and 1925. After four centuries of faith in the existence of rational principles underlying the laws of nature, science suddenly found itself forced to revise its own foundations. The aspect that most disturbed consciences, lulled by the reassuring certainties of the past, was the disconcerting duality of quantum theory. On the one hand, there was abundant experimental evidence that light behaved like a wave, with a lot of interference and diffraction. As we have already seen in detail, the wave hypothesis is the only one able to account for the data obtained from the double slit experiment.

On the other hand, an equally abundant harvest of experiences strongly demonstrated the particle nature of light - and we saw it in the previous one with black body radiation, photoelectric effect and Compton effect. The logical conclusion to which these experiments led was one and only one: light of any color, therefore of any wavelength, was constituted by a flow of particles, all moving in vacuum at the same speed. Each one had its own momentum, a quantity that in Newtonian physics was given by the product of mass for speed and that for photons is equal to the energy divided by c. The momentum is important (as can be testified by whoever passed in front of a speed camera with the car), because its total in a system is preserved, that is it does not change even after several impacts. In the classic case is known the example of two billiard balls colliding: even if the speeds change, the sum of the momentum before and after the collision remains constant. Compton's experiment has shown that this conservation is also valid for how many of light, which in this regard behave like cars and other macroscopic objects.

We should stop for a moment to clarify the difference between waves and particles. First of all, the second ones are discrete. Take two glasses, one full of water and one full of fine sand. Both substances change shape and can be poured, so that, on a not too thorough examination, they seem to share the same properties. But the liquid is continuous, smooth, while the sand is made up of discreet, accounting grains. A teaspoon of water contains a certain volume of liquid, a teaspoon of sand can be quantified in the number of grains. Quantum mechanics re-evaluates discrete quantities and integers, in what would seem a return to Pythagorean theories. A particle, in every instant, has a defined position and moves along a certain trajectory, unlike a wave, which is "smeared" in space. The particles, moreover, have a certain energy and momentum, which can transfer to other particles in collisions. By definition, a particle cannot be a wave and vice versa.

Back to us. Physicists of the twenties were bewildered in front of that strange beast, half particle and half wave, which some called wavicle (wave, wave, and particle, particle contraction). Despite the consolidated evidence in favor of wave nature, experiment after experiment the photons turned out to be concrete objects, able to collide with each other and with electrons. Atoms emitted one when they came out from an excited state, releasing the same amount of energy, $E=hf$, brought by the photon itself. The story took an even more surprising turn with the entry of a young French physicist, the aristocrat Louis-Cesar-Victor-Maurice de Broglie, and his memorable doctoral thesis.

The de Broglie, whose ranks included only high-ranking officers, diplomats or politicians, were very hostile to Louis' inclinations. The old duke, his grandfather, called science "an old woman who seeks the admiration of young men. So the scion, for the sake of compromise, undertook his studies to become a naval officer, but continued to experiment in his spare time, thanks to the laboratory he had installed in the ancestral mansion. In the navy he made a name for himself as an expert in broadcasting and after the death of the old duke he was allowed to take leave to devote himself full-time to his true passion.

De Broglie had thought at length about Einstein's doubts about the photoelectric effect, which was incompatible with the undulating nature of light and which corroborated the photon hypothesis. While rereading the work of the great scientist, the young Frenchman came up with a very unorthodox idea. If light, which would seem to be a wave, exhibits particle-like behavior, perhaps in nature the opposite can also happen. Maybe particles, all particles, show in certain occasions wave-like behaviors. In de Broglie's words: "[Bohr atom theory] led me to hypothesize that also electrons could be considered not only particles, but also objects to which it was possible to assign a frequency, that is a wave property".

Some years before, a doctoral student who had chosen this risky hypothesis for his thesis would have been forced to move to the theology faculty of some obscure university of Molvania Citeriore. But in 1924 everything was possible, and de Broglie had a very special fan. The great Albert Einstein was called by his perplexed Parisian colleagues as external consultant to examine the candidate's thesis, which he found very interesting (perhaps he also thought "but why didn't I have this idea?"). In his report to the Parisian commission, the Master wrote: "De Broglie has lifted a flap of the great veil". The young Frenchman not only obtained the title, but a few years later he was even awarded the Nobel Prize for physics, thanks to the theory presented in his thesis. His greatest success was having found a relationship, modeled on Planck's one, between the classical momentum of an electron (mass for speed) and the wavelength of the corresponding wave. But wave of what? An electron is a particle, for God's sake, where is the wave? De Broglie was talking about "a mysterious phenomenon with periodicity characters" that happened inside the particle itself. It seems unclear, but he was convinced about it. And although his interpretation was smoky, the underlying idea was brilliant.

In 1927 two American physicists working at the prestigious Bell Laboratories of AT&T, New Jersey, were studying the properties of vacuum tubes bombarding with electron flows various types of crystals. The results were quite strange: the electrons were coming out from the crystals according to preferred directions and seemed to ignore others. The Bell Lab researchers were not able to do it, until they found out about de Broglie's crazy hypothesis. Seen in this new point of view, their experiment was only a complex version of Young's one, with the double slit, and the behavior of the electrons showed a well known property of the waves, that is the diffraction! The results would have made sense if it was assumed that the wavelength of the electrons was really related to their momentum, just as de Broglie predicted. The regular lattice of the atoms in the crystals was the equivalent of the cracks in Young's experiment, that was more than a century before. This fundamental discovery of "electronic diffraction" corroborated de Broglie's thesis: electrons are particles that behave like waves, and it is also quite easy to verify it.

We will come back in a little while on diffraction issue, remaking our now familiar double slit experiment with electrons, that will give us an even more disconcerting result. Here we only observe that this property is responsible for the fact that the various materials behave as conductors, insulators or semiconductors, and it is at the base of transistors invention. Now we have to get acquainted with another protagonist - maybe the real superhero of the quantum revolution.

A strange mathematics

Werner Heisenberg (1901-1976) was the prince of the theorists, so disinterested in laboratory practice that he risked flunking his thesis at the University of Munich because he did not know how batteries worked. Fortunately for him and physics as a whole, he was also promoted. There were other not easy moments in his life. During the First World War, while his father was at the front as a soldier, the scarcity of food and fuel in the city was such that schools and universities were often forced to suspend classes. And in the summer of 1918 young Werner, weakened and undernourished, was forced together with other students to help the farmers on a Bavarian farm harvest.

With the end of the war, in the first years of the twenties we find him in the shoes of the young prodigy: pianist of high level, poured in the classical languages, skillful skier and alpinist, as well as mathematician of rank lent to the physics. During the lessons of the old teacher Arnold Sommerfeld, he met another promising young man, Wolfgang Pauli, who would later become his closest collaborator and his fiercest critic. In 1922 Sommerfeld took the 21-year-old Heisenberg to Göttingen, then the beacon of European science, to attend a series of lectures dedicated to the nascent quantum atomic physics, given by Niels Bohr himself. On that occasion the young researcher, not at all intimidated, dared to counter some statements of the guru and challenge at the root his theoretical model. After this first confrontation, however, between the two was born a long and fruitful collaboration, marked by mutual admiration.

From that moment Heisenberg devoted himself body and soul to the enigmas of quantum mechanics. In 1924 he spent some time in Copenhagen to work directly with Bohr on the problems of radiation emission and absorption. There he learned to appreciate the "philosophical attitude" (in Pauli's words) of the great Danish physicist. Frustrated by the difficulties to make concrete the atomic model of Bohr, with its orbits put in that way who knows how, the young man was convinced that there must be something wrong at the root. The more he thought about it, the more it seemed to him that those simple, almost circular orbits were a surplus, a purely intellectual construct. To get rid of them, he began to think that the very idea of orbit was a Newtonian residue that had to be done without.

The young Werner imposed himself a fierce doctrine: no model had to be based on classical physics (so no miniature solar systems, even if they are so cute to draw). The way to salvation was not intuition or aesthetics, but mathematical rigor. Another of his conceptual dikats was the renunciation of all entities (such as orbits, in fact) that could not be measured directly.

Measurable in the atoms were the spectral lines, witness of the emission or absorption of photons by the atoms as a result of jumping between the electron levels. So it was to those net, visible and verifiable lines corresponding to the inaccessible subatomic world, that Heisenberg turned his attention. To solve

95

this diabolically complicated problem, and to find relief from hay fever, in 1925 he retired to Helgoland, remote island in the North Sea.

His starting point was the so-called "correspondence principle", enunciated by Bohr, according to which quantum laws had to be transformed without problems into the corresponding classical laws when applied to sufficiently large systems. But how big? Enough to allow to neglect the Planck constant h in the relative equations. A typical object of the atomic world has mass equal to 10-27 kg; let's consider that a grain of dust barely visible to the naked eye can weigh 10-7 kg: very little, but it is still greater by a factor 100000000000000, that is 1020, one followed by twenty zeros. So the atmospheric dust is clearly in the domain of classical physics: it is a macroscopic object and its motion is not affected by the presence of factors dependent on Planck's constant. The basic quantum laws apply naturally to the phenomena of the atomic and subatomic world, while it loses sense to use them to describe phenomena related to aggregates larger than atoms, as the dimensions grow and quantum physics gives way to the classical laws of Newton and Maxwell. The foundation of this principle (as we will repeat several times) is that the strange and unpublished quantum effects "correspond" directly to the classical concepts of physics as you leave the atomic field to enter the macroscopic one.

Driven by Bohr's ideas, Heisenberg redefined in quantum field the most banal notions of classical physics, as the position and velocity of an electron, so that they were in correspondence with the Newtonian equivalents. But he soon realized that his efforts of reconciliation between two worlds led to the birth of a new, and bizarre, "algebra of physics".

We all learned in school the so-called commutative property of multiplication, that is, the fact that, given any two numbers a and b, their product does not change if we exchange them between them; in symbols: $a \times b = b \times a$. It is obvious, for example, that $3 \times 4 = 4 \times 3 = 12$. In Heisenberg's time, however, the existence of abstract numerical systems in which the commutative property does not always apply and it is not said that $a \times b$ is equal to $b \times a$. To well think about it, examples of non commutative operations are found also in nature. A classic case are rotations and tilts (try to perform two different rotations on an object like a book, and you will find examples where the order in which they happen is important).

Heisenberg had not studied in depth the most advanced frontiers of pure mathematics of his time, but he could avail himself of the help of more experienced colleagues, who immediately recognized the type of algebra contained in his definitions: they were nothing but multiplications of matrices with complex values. So-called "matrix algebra" was an exotic branch of mathematics, known for about sixty years, which was used to treat objects formed by rows and columns of numbers: matrices. Matrix algebra applied to Heisenberg's formalism (called matrix mechanics) led to the first concrete arrangement of quantum physics. His calculations brought to sensible results for the energies of states and atomic transitions, that is electron level jumps.

When the matrix mechanics was applied not only to the hydrogen atom case, but also to other simple microscopic systems, it was discovered that it worked wonderfully: the solutions obtained theoretically agreed with the experimental data. And from those strange manipulations of matrices came out also a revolutionary concept.

The first steps of the uncertainty principle

The main consequence of non-commutationality turned out to be this. If we indicate with x the position along an axis and p the momentum, always along the same axis, of a particle, the fact that xp is not equal to px implies that the two values cannot be measured simultaneously in a defined and precise way. In other words, if we get the exact position of a particle we disrupt the system in such a way that it is no longer possible to know its momentum, and vice versa. The cause of this is not technological, it is not our instruments that are inaccurate: it is nature that is made this way.

In the formalism of matrix mechanics we can express this idea in a concise way, which has always driven the philosophers of science crazy: "The uncertainty relative to the position of a particle, indicated with Δx, and that relative to the quantity of motion, Δp, are linked by the relation: $\Delta x \Delta p \geq \hbar/2$, where ħ=h/2π". Said in words: "the product of the uncertainties relative to the position and the momentum of a particle is always greater or equal by a number equal to the Planck constant divided by four times pi". This implies that if we measure position with great precision, thus making Δx as small as possible, we automatically make Δp arbitrarily large, and vice versa. You just can't have it all in life: we have to give up knowing exactly either the position, or the momentum.

Starting from this principle we can also deduce the stability of Bohr atom, that is to demonstrate the existence of a fundamental state, a lower orbit under which electron cannot descend, as it happens in Newtonian mechanics. If the electron would get closer and closer to the nucleus until it would hit us, the uncertainty about its position would be less and less, that is as scientists say Δx "would tend to zero". For Heisenberg principle Δp would become arbitrarily large, that is the electron energy would grow more and more. It is shown that there is a state of equilibrium where the electron is "enough" well located, with Δx different from zero, and where the energy is the minimum possible, given the corresponding Δp value.

The physical ratio of uncertainty principle is easier to understand if we put ourselves in another order of reasoning, à la Schrödinger, and we examine a property (not quantum) of electromagnetic waves, well known in the telecommunication field. Well yes, we are going back to the waves. The matrix mechanics seemed at first sight the only rigorous way to penetrate in the meanders of the atomic world. But luckily, while physicists were about to become algebra experts, in 1926 another, more attractive solution to the problem appeared.

The most beautiful equation in history

We already met Erwin Schrödinger in chapter 1. As you remember, at one point he took a vacation in Switzerland to study in peace, and the result of this period was an equation, Schrödinger's equation, which brought considerable clarity to the quantum world.

Why is it so important? Let's go back to Newton's first law, the F=ma that governs the motion of apples, planets and all macroscopic objects. It tells us that the force F applied to an object of mass m produces an acceleration (i.e. a change of speed) to and that these three quantities are linked by the relationship written above. Solving this equation allows us to know the state of a body, for example a tennis ball, at every moment. The important thing, in general, is to know F, from which the position x and the velocity v at the instant t are derived. The relationships between these quantities are established by differential equations, that use concepts of infinitesimal analysis (invented by Newton himself) and that sometimes are difficult to solve (for example when the system is composed by many bodies). The shape of these equations is however rather simple: they are the calculations and the applications to become complicated.

Newton amazed the world demonstrating that joining the law of universal gravitation to that one of motion, applied to the force of gravity, the simple elliptical orbits and the laws of planetary motion that Kepler had enunciated for the solar system were obtained. The same equation is able to describe the motions of the Moon, of an apple falling from the tree and of a rocket fired in orbit. This equation, however, is not able to solve explicitly if four or more bodies are involved, all subject to gravitational interaction; in this case it is necessary to proceed by approximations and/or with the help of numerical methods (thanks to calculators). It's a nice case: at the base of the laws of nature there is an apparently simple formula, but it reflects the incredible complexity of our world. Schrödinger's equation is the quantum version of F=ma. If we solve it, however, we do not find ourselves with the position and velocity values of the particles, as in the Newtonian case.

On that vacation in December 1925 Schrödinger brought with him not only his mistress, but also a copy of de Broglie's doctoral thesis. Very few, at the time, had noticed the ideas of the Frenchman, but after Schrödinger's reading things changed quickly. In March 1926, this forty years old professor from the University of Zurich, who until then had not had a particularly brilliant career and that by the standards of the young physicist of that time was almost decrepit, made known to the world his equation, that dealt with the motion of electrons in terms of waves, based on de Broglie's thesis. For his colleagues he was much more digestible than the cold abstractions of matrix mechanics. In Schrödinger's equation a new fundamental quantity appeared, the wave function, indicated with Ψ, that represents its solution.

Since long before the official birth of quantum mechanics, physicists were used to treat (classically) various cases of material waves in the continuous, such as sound waves that propagate in the air. Let's see an example with sound. The quantity that interests us is the pressure exerted by the wave in the air, which we indicate with $\Psi(x,t)$. From the mathematical point of view this is a "function", a recipe that provides the value of the wave pressure (intended as a variation of standard atmospheric pressure) at each point x of space and at each instant t. The solutions of the relative classical equation naturally describe a wave that "travels" in space and time, "perturbing" the motion of air particles (or water, or an electromagnetic field or other). The waves of the sea, tsunamis and beautiful company are all forms allowed by these equations, which are of the "differential" type: they involve quantities that change, and to understand them you need to know the mathematical analysis. The "wave equation" is a type of differential equation that if solved gives us the "wave function" $\Psi(x,t)$ - in our example the air pressure that varies in space and time when a sound wave passes.

Thanks to de Broglie's ideas, Schrödinger immediately understood that Heisenberg's complex technicalities could be rewritten in order to obtain relations very similar to the old equations of classical physics, in particular those of waves. From the formal point of view, a quantum particle was described by the function $\Psi(x,t)$, which Schrödinger himself called "wave function". With this interpretation and applying the principles of quantum physics, i.e. solving Schrödinger's equation, it was possible in principle to calculate the wave function of each particle then known, in almost all cases. The problem was that nobody had any idea what this quantity represented.

As a consequence of the introduction of Ψ, we can no longer say that "at instant t the particle is in x"; instead we have to say that "the motion of the particle is represented by the function $\Psi(x,t)$, which gives the amplitude Ψ at time t in point x". The precise position is no longer known. If we see that Ψ is particularly large at a point x and almost nothing elsewhere, we can say that the particle is "about in position x". Waves are objects scattered in space, and so is the wave function. We observe that these are reasoning with hindsight, because in the years we are considering no one, including Schrödinger, had very clear ideas about the true nature of the wave function.

Here, however, is a twist, which is one of the most surprising aspects of quantum mechanics. Schrödinger realized that his wave function was, as one would expect from a wave, continuous in space and time, but that to make the accounts come back he had to assume different numbers as values from the real ones. And this is a big difference with normal waves, mechanical or electromagnetic, where values are always real. For example, we can say that the crest of an oceanic wave rises from the average sea level for 2 meters (and therefore we must expose the red flag on the beach); or even worse that a tsunami 10 meters high is coming, and therefore we must evacuate the coastal areas in a hurry. These are real values, concrete, measurable with various instruments, and we all understand their meaning.

The quantum wave function, on the contrary, assumes values in the field of so-called "complex numbers".15 For example, it may happen that at point x the amplitude is equal to a "stuff" that is written 0.3+0.5i, where $i=\sqrt{-1}$. In other words, the number i multiplied by itself gives the result -1. An object like the one written above, formed by a real number added to another real number multiplied by i, is defined as a complex number. Schrödinger's equation always involves the presence of i, a number that plays a fundamental role in the equation itself, which is why the wave function assumes complex values.16

This mathematical complication is an unavoidable step on the road to quantum physics and is another indication that the wave function of a particle is not directly measurable: after all, real numbers are always obtained in experiments. In Schrödinger's vision, an electron is a wave for all purposes, not different from a sound or marine wave. But how is it possible, since a particle must be located in a defined point and cannot occupy entire portions of space? The trick is to superimpose various waves in such a way that they are erased almost everywhere except at the point we are interested in. A combination of waves, therefore, can represent an object well located in space, which we would be tempted to call "particle" and that comes up every time the sum of the waves gives rise to a particular concentration in a point. In this sense a particle is an "anomalous wave", similar to that phenomenon caused at sea by the overlapping of waves, which creates a great perturbation capable of overturning boats.

An eternal adolescent

At what point is quantum theory after the discoveries of Heisenberg, Schrödinger, Bohr, Born and colleagues? There are the probabilistic wave functions on the one hand and the uncertainty principle on the other, which allows to maintain the particle model. The crisis of duality "a bit wave a bit particle" seems to be solved: electrons and photons are particles, whose behavior is described by probabilistic waves. As waves they are subject to interference phenomena, bringing the docile particles to appear where they are expected to do so, obeying the wave function. How they get there is not a problem that makes sense to pose. This is what they say in Copenhagen. The price to pay for success is the intrusion in physics of probability and various quantum oddities.

The idea that nature (or God) plays dice with subatomic matter did not appeal to Einstein, Schrödinger, de Broglie, Planck and many others. Einstein in particular was convinced that quantum mechanics was only a stage, a provisional theory that would sooner or later be replaced by another, deterministic and causal one. In the second part of his career the great physicist made several ingenious attempts to circumvent the problem of uncertainty, but his efforts were frustrated one after the other by Bohr - supposedly to his malicious satisfaction.

We must therefore close the chapter suspended between the triumphs of theory and a certain feeling of uneasiness. At the end of the twenties quantum mechanics was now an adult science, but still susceptible to growth: it would be deeply revised several times, until the forties.

Chapter 6
Quantum

As if to confirm its supernatural appearance, the quantum theory of Heisenberg and Schrödinger literally worked miracles. The hydrogen atom model was clarified without the need of Keplerian conceptual crutches: the orbits were replaced by "orbital", sons of the new and indeterminate wave functions. The new quantum mechanics turned out to be a formidable tool in the hands of physicists, who became more and more good to apply Schrödinger's equation to various fields and atomic and subatomic systems, of increasing complexity. As Heinz Pagels said, "the theory released the intellectual energies of thousands of young researchers in all industrialized nations. In no other occasion has a series of scientific ideas had such fundamental consequences on technological development; its applications continue to shape the political and social history of our civilization".

But when we say that a theory or model "works", what exactly do we mean? That it is mathematically able to make predictions about some natural phenomenon, comparable with experimental data. If the predictions and measurements accumulated in our experiences collimate, then the theory works "ex post", i.e. it explains why a certain fact happens, which was previously unknown to us.

For example we could ask ourselves what happens to throw two objects of different mass from a high point, say the tower of Pisa. Galileo's demonstration and all the experiments performed afterwards show that, unless small corrections due to air resistance, two serious objects of different mass falling from the same height reach the ground at the same time. This is one hundred percent true in the absence of air, as has been spectacularly demonstrated on the Moon live on television: the feather and hammer dropped by an astronaut arrived at exactly the same time.2 The original and profound theory that has been confirmed in this case is Newtonian universal gravitation, combined with its laws of motion. Putting together the relative equations, we can predict what will be the behavior of a falling body subject to the force of gravity and calculate how long it will take to reach the ground. It is child's play to verify that two objects of different masses dropped from the same height must reach the ground at the same time (if we neglect air resistance).

But a good theory must also make us able to predict the evolution of phenomena not yet observed. When the ECHO satellite was launched in 1958, for example, gravitation and Newtonian laws of motion were used to calculate in advance what trajectory it would follow, note the thrust force and other important corrective factors, such as wind speed and earth rotation. The predictive power of a law depends, of course, on how much control you can exercise over the various factors involved. Under every point of view, Newton's theory has proved to be of extraordinary success, both in ex post verifications and in the predictive field, when applied to its wide range of relevance: not too high speeds (much smaller than light) and not too small scales (much larger than atomic).

Newton does not write e-mail

Let's ask ourselves now if quantum mechanics is able to explain (ex post) the world around us and if it can be used to predict the existence of phenomena not yet observed - thus becoming indispensable in the discovery of new and useful applications. The answer to both questions is a convinced yes. The quantum theory has passed countless experimental tests, in both senses. It is grafted on the theories that preceded it, Newtonian mechanics and Maxwell's electromagnetism, every time the quantum trademark, i.e. Planck's famous constant h, is not so small that it can be ignored in the calculations. This happens when masses, dimensions and time scales of objects and events are comparable to those of the atomic world. And since everything is made up of atoms, we should not be surprised that these phenomena sometimes raise their heads and make their presence felt even in the macroscopic world, where humans and their measuring instruments are.

In this chapter we will explore the applications of this strange theory, which will seem related to witchcraft. We will be able to explain all the chemistry, from the periodic table of the elements to the forces that hold together the molecules of the compounds, of which there are billions of types. We will then move on to verify how much quantum physics affects practically every aspect of our life. If it is true that God plays dice with the universe, he has however managed to control the results of the game in order to give us the transistor, the tunnel diode, lasers, X-rays, synchrotron light, radioactive markers, scanning tunnel effect microscopes, superconductors, positron emission tomography, superfluids, nuclear reactors, atomic bombs, magnetic resonance imaging and microchips, just to give a few examples. You probably do not have superconductors or scanning microscopes, but you certainly have hundreds of millions of transistors scattered around the house. Your life is touched in a thousand ways by technologies possible thanks to quantum physics. If we had a strictly Newtonian universe, we wouldn't be able to surf the Internet, we wouldn't know what software is and we wouldn't have seen the battles between Steve Jobs and Bill Gates (or rather, they would have been billionaire rivals in another sector, such as railways). Perhaps we would have saved ourselves some problems that afflict our time, but for sure we would not have the tools to solve many others.

The consequences in other scientific fields, beyond the boundaries of physics, are equally profound. Erwin Schrödinger, to whom we owe the elegant equation that governs the entire quantum world, wrote in 1944 a prophetic book entitled What is life,4 in which he made a hypothesis about the transmission of genetic information. The young James Watson read this remarkable work and was stimulated to investigate the nature of genes. The rest of the story is well known: together with Francis Crick, in the 1950s Watson discovered the double helix of DNA, starting the revolution of molecular biology and, later, the unscrupulous genetic engineering of our times. Without the quantum revolution we would not have been able to understand the structure of the simplest molecules, let alone the DNA, which is the basis of all life.5 Moving into more frontier and speculative areas, those who could offer the solution to problems such as the nature of the mind, consciousness and self-perception, or at least this is what some daredevil theoretical physicists who dare face the field of cognitive sciences claim.

Quantum mechanics continues to shed light on chemical phenomena to this day. In 1998, for example, the Nobel Prize in Chemistry was awarded to two physicists, Walter Kohn and John Pople, for discovering powerful computational techniques for solving quantum equations that describe the shape and interactions of molecules. Astrophysics, nuclear engineering, cryptography, materials science, electronics: these branches of knowledge and others, including chemistry, biology, biochemistry and so on, would be impoverished without quanta. What we call computing would probably be little more than the design of archives for paper records. What would this discipline do without Heisenberg's uncertainty and Born's probabilities?

Without quantum we would not have been able to really understand the structure and properties of the chemical elements, which had been sitting well in the periodic table for half a century. It is the elements, their reactions and their combinations that give life to everything around us and to life itself.

A game with Dmitri Mendeleev

Chemistry was a serious and vital science, like physics, long before quantum theory entered the scene. It was through the chemical research of John Dalton that in 1803 the reality of atoms was confirmed, and Michael Faraday's experiments led to the discovery of their electrical properties. But nobody understood how things really were. Quantum physics provided chemistry with a sophisticated and rational model capable of explaining the detailed structure and behavior of atoms, as well as a formalism to understand the properties of molecules and realistically predict their formation. All these successes were possible thanks to the probabilistic nature of the theory.

We know, chemistry is not a popular subject, although it is the basis of much of modern technology. All those symbols and numbers written below confuse ideas. But we are convinced that if you let yourself be led in this chapter to explore the logic behind the discipline, you will be won over. The discovery of the mysteries of the atom is one of the most compelling mystery novels in the history of mankind.

The study of chemistry, as everyone knows, starts from the periodic table of the elements, which adorns the walls of hundreds of thousands of classrooms around the world. His invention was a true scientific feat, accomplished largely by the surprisingly prolific Dmitri Ivanovič Mendeleev (1834-1907). A leading figure of czarist Russia, Mendeleev was a great scholar, able to write four hundred books and articles, but he was also interested in the practical applications of his work, so much so that he left contributions on subjects such as the use of fertilizers, cheese production in cooperatives, standardization of weights and measures, customs tariffs in Russia and shipbuilding. Politically radical, he divorced his wife to marry a young student at the art institute. Judging by period photos, he liked to keep his hair long.

Mendeleev's diagram comes from ordering the elements by increasing atomic weight. We observe that by "element" we mean a simple substance, consisting of atoms all of the same type. A block of graphite and a diamond are made of the same substance, carbon, even if the atoms have different structure: in one case they give rise to a dark and useful material to make pencils, in the other to objects useful to make themselves beautiful in the eyes of the girlfriend or to pierce the hardest of metals. Vice versa water is not an element but a chemical compound, because it is composed of oxygen and hydrogen atoms held together by electrical forces. Even the compound molecules obey Schrödinger's equation.

The "atomic weight" that we mentioned before is nothing but the characteristic mass of each atom. All atoms of the same substance, let's say oxygen, have the same mass. Ditto for nitrogen atoms, which are a little less heavy than oxygen atoms. There are very light substances, like hydrogen, the lightest of all, and other very heavy substances, like uranium, hundreds of times more massive than hydrogen. The atomic mass is measured for convenience with a special unit of measure and it is indicated with the letter M,8 but here it is not important to go into details of the single values. Rather we are interested in the list of elements in increasing order of atomic weight. Mendeleev realized that the position of an element in this list had a clear correspondence with its chemical properties: it was the key to penetrate the mysteries of matter.

Mr. Pauli enters the scene

Physical systems tend to organize themselves so that they remain in the state of least energy. In atoms, quantum rules and Schrödinger's equation provide the allowed configurations in which electrons can move, the orbital ones, each of which has its own energy level. The last step to unravel all the mysteries of the atom starts from here, and it is an extraordinary and surprising discovery: each orbital has space only for a maximum of two electrons! If this is not the case, the physical world would be very different.

This is where the genius of Wolfgang Pauli comes in, this irascible and legendary scientist who represented in a certain way the consciousness of his generation, the man who terrorized colleagues and students, that sometimes signed himself "The Wrath of God" and that we will hear more often about (see also note 1 in chapter 5).

To avoid electron piles in s1, in 1925 Pauli hypothesized that the so-called exclusion principle was valid, according to which two electrons inside an atom can never be in the same quantum state simultaneously. Thanks to the exclusion principle there is a criterion to put the particles in the right place as you go up to heavier and heavier atoms. By the way, the same principle is what prevents us to pass through walls, because it assures us that the electrons in our body cannot be in the same state as the ones in the wall and they must remain separated by large spaces, like houses in the Big Prairies.

Professor Pauli was a short, chubby, creative and hypercritical gentleman, endowed with a sarcastic spirit that was terror and delight of his colleagues. He certainly did not lack modesty, since he wrote an article as a teenager in which he explained the theory of relativity to physicists in a convincing way. His career was punctuated by lightning-fast jokes, among which we remember for example: "Ah, so young and already so unknown!", "This article does not even have the honor of being wrong", "Your first formula is wrong, but in any case the second is not derived from the first", "I am not bothered by the fact that you are slow in

115

understanding, but that you write articles faster than you think". To be the subject of one of these arrows was certainly an experience that could reduce anyone's size.

An author who remained anonymous wrote this poem about Pauli, as it is reported by George Gamow in his book Thirty years that upset physics:

The principle of exclusion was one of Pauli's greatest scientific achievements. In practice he gave us back chemistry, allowing us to understand why the periodic table of elements is made that way. Yet in its basic formulation is very simple: never two electrons in the same quantum state. It is not done, verboten! This little rule guides us in the construction of bigger and bigger atoms and in the understanding of their chemical properties.

We repeat the two golden rules, by Pauli, that we must follow as we go along the periodic table: 1) electrons must always occupy different quantum states and 2) electrons must be configured to have the minimum possible energy. This second rule, incidentally, applies in other areas and it explains why bodies subject to gravity fall down: an object on the ground has less energy than one on the fourteenth floor. But let's go back to helium. We said that the two electrons in s1 are consistent with the experimental data. Is this not a violation of the exclusion principle ? Actually no, because thanks to another great idea of Pauli, maybe the most brilliant one, the spin comes on the scene (in addition to what we will say now, for a deeper discussion of this topic see Appendix).

The electrons, in a certain way, rotate on themselves in a relentless way, like microscopic spinning tops. This rotation, from quantum point of view, can happen in two ways, that are called up (up) and down (down). That is why two electrons can easily stay in the same orbital 1s and respect Pauli's diktat: as long as they have opposite spin they can find themselves in different quantum states. Here it is done. Now however we have exhausted 1s, we cannot overload it with a third electron.

The helium atom saturates the orbital 1s and it is ok, because there is no more space available: the two electrons are sitting tight and nice and quiet. The consequence of this structure is just the chemical inactivity of helium, that has no desire to interact with other atoms. Hydrogen instead has only one electron in 1s and it is hospitable to other particles that want to join it, as long as they have the opposite spin; in fact the arrival of an electron from another atom (as we will see soon) is the way hydrogen creates a bond with other elements.16 In chemistry language, the hydrogen orbital is called "incomplete" (or even that its electron is "odd"), while the helium one is "complete", because it has the maximum number of electrons expected: two, of opposite spin. The chemistry of these two elements, therefore, is different like day from night.

Now we are ready to face lithium and its three electrons. Where do we put them ? Two can fill 1s, as for helium, and the third one we put it necessarily in the lowest energy level among those available. As you can see in figure 6.5, we have four choices: 2s,

2px, 2py and 2pcs. The chemical properties of lithium depend only on this last electron, because the ones in 1s complete it, making it inert. We know that 2s is slightly less energetic than the various 2p, so the choice is forced and we put it there. In the end it turns out that lithium is chemically almost identical to hydrogen, as it too has only one active electron in the outer orbital (first in 1s, now in 2s). Slowly we are revealing the secret code that leads to the periodic table formation.

The heavier atoms are formed filling one after the other the orbital 2s and 2p. Beryllium has the fourth electron in one of the 2p (because of a minimum repulsive action of the other electron in 2s). All the 2p, that as already said are drop shaped, have the same energy, so an electron can end up in a quantum "mixture" of these three states. Let us always remember, while we continue to fill the orbital, that each one contains at most two electrons, one with spin up and the other with spin down. The Sequence of the heaviest atoms proceeds smoothly: beryllium (Z=4), boron (Z=5), carbon (Z=6), nitrogen (Z=7), oxygen (Z=8), fluorine (Z=9) and neon (Z=10). Neon is exactly eight steps after helium. What is the situation at this point? 1s, 2s, 2px, 2py and 2pcs are all complete, each one with its good two electrons, so we find ourselves in the case of helium, with the ten neon electrons that are quiet in their orbital. Here is explained the reason of the magic number eight and the periodicity in Mendeleev's table.

Summing up, hydrogen and lithium are chemically similar because they both have only one electron in the outer orbital (respectively 1s and 2s). Helium and neon are chemically similar because all their electrons are in complete orbitals (respectively 1s and 1s, 2s, 2px, 2py and 2pz), which implies stability and chemical non reactivity. It is the incomplete levels, in fact, to stimulate the atoms to activity. The mystery of the suspects in the American-style confrontation that gathered in secret bands, observed for the first time by Mendeleev, is almost completely cleared up.

Now it is up to sodium, Z=11, with eleven positive charges in the nucleus and eleven electrons that somehow we have to fix. We have already seen that the first ten complete all the first five orbitals, so we must turn to 3s and place there the solitary electron. Voilà: sodium is chemically similar to hydrogen and lithium, because all three of them have only one electron in the outer orbital, that is s type. Then there is magnesium, that adds to the electron in 3s another one shared (in quantum sense) between 3px, 3py and 3pcs. Continuing to fill 3s and 3p, we realize that the configurations exactly replicate the ones already seen for 2s and 2p; after other eight steps we arrive to the argo, another noble and inert gas that has all the complete orbitals: 1s, 2s, 2p, 3s and 3p - all contain their good two opposite spin electrons. The third row of the periodic table replicates exactly the second one, because the s and p orbitals of its atoms fill in the same way.

In the fourth row, however, things change. We start quietly with 4s and 4p but then we come across the 3d. The orbits of this type correspond to even higher order solutions of Schrödinger's equation and they fill up in a more complicated way, because at this point there are a lot of electrons. So far we have neglected this aspect, but the negatively charged particles interact with each other, rejecting each other because of the electric force, which complicates the calculations a lot. It is the equivalent of the Newtonian problem of n bodies, that is similar to the situation in which you have to find the motions of a solar system whose planets are close enough to each other to feel the respective gravitational influence. The details of how it is possible to reach a solution are complex and are not relevant for what we will say here, but it is enough to know that everything works in the end. The 3d orbital are mixed with the 4p, so you can find place for a maximum of ten electrons before they complete. That is why the period eight changes in eighteen (18=8+10) and then it changes again for similar reasons in thirty-two. The physical bases of the chemical behavior of the ordinary matter, and therefore also of the substances that allow life, are now clear. The mystery of Mendeleev is no longer such.

to the details.

Chapter 7
Einstein and Bohr

W e have overcome the obstacle of the last chapters and we have come to understand, thanks to Wolfgang Pauli, the intimate essence of chemistry (and therefore of biology) and why we will never be able, with high probability, to pass through a granite table with our hand, which also consists mostly of empty space. The time has come to go even deeper into the sea of quantum mysteries and deal with the fundamental dispute between Niels Bohr and Albert Einstein. We will hear some good things about it.

Scientific creativity, ideally, is an eternal battle between intuition and the need for incontrovertible proof. Today we know that quantum science successfully describes an incredible number of natural phenomena and even has applications with very concrete economic spin-offs. We have also realized that the microscopic, i.e. quantum, world is strange, but strange indeed. Today's physics does not even seem related to the one that has made so much progress from the seventeenth century to the beginning of the twentieth century. There has been a real revolution.

Sometimes scientists (including us), in an attempt to get the results of their research to the general public, resort to metaphors. They are in a sense daughters of the frustration of those who cannot explain what they saw in the laboratory in a "sensible" way, because this would involve a revision of our way of thinking: we must try to understand a world whose direct and daily experience is precluded us. Surely our language is inadequate to describe it, since it has evolved for other purposes. Let's assume that a race of aliens from the planet Zyzzx has collected and analyzed certain macroscopic data from planet Earth and now knows everything about the behavior of the crowds - from stadium games to concerts in the square, from marching armies to mass protests ended with the crowd running away from police charges (which only happens in backward countries, eh). After collecting information for a century, the Zyzzxians have a substantial catalog of collective actions at their disposal, but they know nothing about the abilities and aspirations of men, reasoning, love, passion for music or art, sex, humor. All these individual characteristics are lost in the molasses of collective actions.

The same happens in the microscopic world. If we think about the fact that a flea's cilia hair contains a thousand billion billion atoms, we understand why macroscopic objects, elements of our daily experience, are useless in the understanding of microscopic reality. Like the individuals in the crowd, atoms blend into tangible bodies - though not entirely, as we will see later. So we have two worlds: a classic one, elegantly described by Newton and Maxwell, and a quantum one. Of course, at the end of the day the world is only one, in which quantum theory can successfully deal with atoms and merge into the classical one in the macroscopic case. The equations of Newton and Maxwell are approximations of the quantum ones. We review in a systematic way some disconcerting aspect of these last ones.

Four shocks in a row

1. Let's start with radioactivity and start with one of the most beloved creatures of physicists, the muon. It is a particle that weighs about two hundred times more than an electron, of which it has the same charge. It seems to be punctiform, that is it has negligible size, and it seems to turn on itself. When this heavy copy of the electron was discovered, it caused a bewilderment in the scientific community, so much that the great Isaac Isidor Rabi came out with the famous exclamation "Who ordered this?".[1] But the muon has a basic difference with the electron: it is unstable, that is subject to radioactive decay, and it disintegrates after about two microseconds. For the precision, its "half-life", that is the time in which on average half of the muons of a thousand group disappear, is equal to 2,2 microseconds. This on average, because if we fixate on a single muon (we can also give it a nice name, like Hilda, Moe, Benito or Julia) we do not know when its life will end. The event "decay of the muon X" is random, not deterministic, as if someone rolled a couple of dice and decided the events based on the combinations of

numbers that came out. We must abandon classical determinism and reason in probabilistic terms, to understand the foundations of the new physics.

2. In the same order of ideas we have the phenomenon of partial reflection, which you will remember from chapter 3. It was thought that light was a wave, subject to all the phenomena of other waves like those of the sea, including precisely reflection, until Planck and Einstein discovered quanta, particles that behave like waves. If a quantum of light, that is a photon, is shot against a window, it is reflected or diffracted; in the first case it contributes to give a weak image of who is admiring the exposed clothes, in the second one it illuminates the elegant mannequins. The phenomenon is described mathematically by a wave function, which being a wave can be partially reflected or diffracted. The particles, instead, are discrete, so they must be either reflected or refracted, in total.

3. We now come to the already well-known experiment with the double slit, whose noble history dates back to Thomas Young who denies Newton on the corpuscular theory of light and sanctions the triumph of the undulatory one. We have seen it for photons, but in reality all particles behave in the same way: muons, quarks, W bosons and so on. All of them, if subjected to an experiment similar to Young's one, seem to behave like waves.

Let's see for example the electron case, that as the photon can be emitted from a source and shot against a screen where two slits were made, beyond which we placed a detector screen (with special circuits instead of photocells). We perform the experiment firing the electrons slowly, for example one per hour, in order to be sure that the particles pass one at a time (without "interfering" between them). As we have discovered in chapter 4, repeating the measurements many times in the end we find ourselves with a set of electron locations on the screen that form an interference figure. The single particle also seems to "know" if there are one or two open slits, while we do not even know which way it passed. If we close one of the two slits, the interference figure disappears. And it disappears even if next to the slits we place an instrument that records the electrons passing from that point. Ultimately, the interference figure appears only when our ignorance about the path followed by the single electron is total.

4. As if that was not enough, we have to deal with other disturbing properties. Take for example the spin. Perhaps the most disconcerting aspect of the story is given by the fact that the electron has fractional spin value, equal to 1/2, that is its "angular momentum" is $\hbar/2$ (see Appendix). Moreover, an electron is always aligned to any direction along which we choose to measure its spin, which if we take into account the orientation can be +1/2 or -1/2, or as we said above up (up) or down (down).2 The icing on the cake is this: if we rotate an electron in the space of 360°, its wave function from Ψe becomes -Ψe, that is it changes sign (in the Appendix there is a special paragraph

explaining how this happens). None of this happens for objects in the classical world.

Take for example a drumstick. If the percussionist of a band, in the mood for performances, starts to twirl it between his fingers between shots, thus turning it 360°, the object returns to have exactly the same spatial orientation. If instead of the drumstick there was an electron, after the turn we would find ourselves with a particle of opposite sign. Definitely we are in unfamiliar territories. But does it really happen or is it just mathematical sophistication? As always, we can measure only the probability of an event, the square of the wave function, so how do we know if the minus sign appears or not? And what does "minus sign" mean in this case, what does it have to do with reality? Are they not elucubrations of philosophers contemplating the navel at the expense of public funds for research?

Nein! says Pauli. The minus sign implies that, taken two electrons at random (remember that they are all identical), their joint quantum state must be such to change sign if the two are exchanged. The consequence of all this is the Pauli exclusion principle, the exchange force, the orbital filling, the periodic table, the reason why hydrogen is reactive and helium inert, the whole chemistry. This fact is the basis of the existence of stable matter, conductors, neutron stars, antimatter and about half of U.S. gross domestic product.

But why so much strangeness?

Let's go back to point 1 of the previous paragraph and the dear old muon, elementary particle that weighs two hundred times the electron and lives two millionths of a second, before decaying and transforming into an electron and neutrinos (other elementary particles). Despite these bizarre features, they really exist and at Fermilab we hope one day to build an accelerator that will make them run at high speed.

The decay of muons is basically determined by quantum probability, while Newtonian physics is on the sidelines watching. At the time of its discovery, however, not everyone was willing to throw overboard a concept as beautiful as the "classical determinism", the perfect predictability of phenomena typical of classical physics. Among the various attempts to save the salvageable there was the introduction of the so-called "hidden variables".

Let's imagine that inside the muon is hidden a time bomb, a tiny mechanism with its good little clock connected to a dynamite charge, that blows up the particle, even if we do not know when. The bomb must be, therefore, a mechanical device of Newtonian type but submicroscopic, not observable with our current technologies but still ultimately responsible for the decay: the hands of the clock arrive at noon, and hello muon. If when you create a muon (usually after shocks between other types of particles) the time of detonation is established at random (perhaps in ways related to the creation of the hidden mechanism), then we have replicated in a classic way the apparently indeterministic process that is observed. The small time bomb is an example of hidden variable, name given to various similar devices that could have the important effect of modifying quantum theory in a deterministic sense, to sweep away the "senseless" probability. But as we will see soon, after eighty years of disputes we now know that this attempt has failed and the majority of contemporary physicists now accept the strange quantum logic.

The hidden things

In the 1930s, well before the quarks were discovered, Einstein gave vent to his deep opposition to the Copenhagen interpretation with a series of attempts to transform quantum theory into something more like Newton and Maxwell's old, dear, sensible physics. In 1935, with the collaboration of the two young theoretical physicists Boris Podolsky and Nathan Rosen, he pulled the ace up his sleeve.8 His counter-proposal was based on a mental experiment (Gendankenexperiment, we have already talked about it) that in its intentions would have to demonstrate with great force the clash between the quantum world of probability and the classical world of real objects, with defined properties, and would also establish once and for all where the truth was.

This experiment became famous under the name of "EPR paradox", from the initials of the authors. Its purpose was to demonstrate the incompleteness of quantum mechanics, in the hope that one day a more complete theory would be discovered.

What does it mean to be "complete" or "incomplete" for a theory? In this case an example of "completion" is given by the hidden variables seen before. These entities are exactly what they say they are: unknown factors that influence the course of events and that can (or maybe not) be discovered with a more thorough investigation (remember the example of the time bomb inside the muon). In reality they are common presences in everyday life. If we flip a coin, we know that the two results "head" and "cross" are equally likely. In the history of humanity, since the invention of the coins, this gesture will have been repeated thousands of billions of times (perhaps even by Brutus to decide whether to kill or not Caesar). Everyone agrees that the result is unpredictable, because it is the result of a random process. But are we really sure? Here is where the hidden variables jump out.

One is, just to start with, the force used to throw the coin in the air, and in particular how much of this force results in the vertical motion of the object and how much in its rotation on itself. Other variables are the weight and size of the coin, the direction of the micro-currents of air, the precise angle with which it hits the table when it falls and the nature of the surface of impact (is the table made of wood? Is it covered with a cloth?). In short, there are a lot of hidden variables that influence the final result.

Now let's imagine to build a machine able to throw the coin always with the same force. We always use the same specimen and we perform the experiment in a place protected from currents (maybe under a glass bell where we created the vacuum), making sure that the coin always falls near the center of the table, where we also have control over the elasticity of the surface. After having spent, say, $17963.47 for this contraption, we are ready to turn it on. Go! Let's make five hundred flips and get five hundred times head! We have managed to control all the elusive hidden variables, which are now neither variables nor hidden, and we have defeated the case! Determinism is the master! Newtonian classical physics applies, therefore, to coins as much as to arrows, bullets, tennis balls and planets. The apparent randomness of a coin toss is due to an incomplete theory and a large number of hidden variables, which in principle are all explicable and controllable.

On what other occasions do we see randomness at work in everyday life? Actuarial tables are used to predict how long a certain population will live (of humans, but also of dogs and horses), but the general theory of the longevity of a species is certainly incomplete, because many complex hidden variables remain, including the genetic predisposition to contract certain diseases, the quality of the environment and nutrition, the probability of being hit by an asteroid and many others. In the future perhaps, if we exclude occasional accidents, we will be able to reduce the degree of uncertainty and predict better until we enjoy the company of grandparents or cousins.

Physics has already tamed some theories full of hidden variables. Consider for example the theory of "perfect gases" or "ideals", which provides a mathematical relationship between pressure, temperature and volume of a gas in a closed environment under ordinary environmental conditions. In the experiments we find that increasing the temperature also increases the pressure, while increasing the volume the pressure drops. All this is elegantly summarized in the formula pV=nRT (in words: "the product of pressure by volume is equal to temperature multiplied by a constant R, all multiplied by the number n of gas molecules"). In this case the hidden variables are an enormity, because the gas is formed by a colossal number of molecules. To overcome this obstacle, we define the temperature statistically as the average energy of a molecule, while pressure is the average force with which the molecules hit a fixed area of the walls of the container that contains them. The law of perfect gases, an incomplete time, thanks to statistical methods can be precisely justified by the motions of "hidden" molecules. With similar methods, in 1905 Einstein succeeded in explaining the so-called Brownian motion, i.e. the apparently random movements of dust suspended in water. These "random walks" of the grains were an insoluble mystery before Einstein understood that "hidden" collisions with water molecules came into play.

Perhaps it was because of this precedent that Einstein naturally thought that quantum mechanics was in reality incomplete and that its probabilistic nature was only apparent, the result of the statistical average made on still unknown entities: hidden variables. If it had been possible to unveil this internal complexity, it would have been possible to go back to Newtonian physics, deterministic and to re-enter the classical reality underlying the whole. If for example the photons guarded a hidden mechanism in order to decide if to be reflected or refracted, the randomness of their behavior when they collide with the showcase would be revealed only apparent. Knowing the operation of the mechanism, we would be able to foresee the motions of the particles.

We put immediately the things in clear: all this has never been never been discovered. Some physicists like Einstein were disgusted, on a philosophical level, by the idea that randomness was an intrinsic, fundamental characteristic of our world and they hoped to recreate in some way the Newtonian determinism. If we knew and controlled all the hidden variables, they said, we could design an experiment whose result would be predictable, as the core of determinism claims.

On the contrary, quantum theory in Bohr and Heisenberg's interpretation rejected the existence of hidden variables and embraced instead causality and indeterminacy, as fundamental characteristics of nature, whose effect was explicitly exhibited at microscopic level. If we cannot predict the outcome of an experiment, we certainly cannot predict the future course of events: as a natural philosophy, determinism has failed.

So let's ask ourselves if there is a way to discover the existence of hidden variables. First, however, let us see what Einstein's challenge consisted of.

Bohr's response to EPR

The key to solving the EPR paradox lies in the fact that the two particles A and B, however far apart, were born at the same time from the same event and are therefore related in an entanglement. Their positions, momentum, spin and so on are always indefinite, but whatever value they assume, they always remain linked to each other. If for example we obtain a precise number for the speed of A, we know that the speed of B is the same, only opposite in direction; same for position and spin. With the act of measuring we make a wave function collapse that until then included in itself all possible values for the properties of A and B. Thanks to entanglement, however, what we learn in our laboratory on Earth allows us to know the same things about a particle that could be on Rigel 3, without touching it, observing it or interfering with it in any way. Even B's wave function collapsed at the same time, even if the particle is sailing light years away.

All this does not refute in any concrete way Heisenberg's uncertainty principle, because when we measure the momentum of A we continue to disturb its position irreparably. EPR's objection focused on the fact that a body must have precise values of momentum and position, even if we are not able to measure them together. How did you eventually decide to replicate Bohr? How did he counterattack?

After weeks of passion, the Master came to the conclusion that the problem did not exist. The ability to predict the speed of B through the measurement of A does not imply at all that B has such speed: until we measure it, there is really no point in making assumptions of this kind. Similarly for the position, which makes no sense to talk about before measurement. Bohr, to which Pauli and other colleagues later added Pauli and others, argued in practice this: poor Einstein did not get rid of the classical obsession with the properties of bodies. In reality, no one can know whether or not such an object has certain properties until we disturb it with measurement. And something that cannot be known may well not exist. You can't count the angels that can balance on the head of a pin, so they may not exist either. The principle of location in all this is not violated: you will never be able to send an instant message of good wishes from Earth to Rigel 3 if you have forgotten your wedding anniversary.

On one occasion Bohr went so far as to compare the quantum revolution with the one triggered by Einstein, relativity, after which space and time found themselves with new and bizarre qualities. The majority of physicists, however, agreed that the former had much more radical effects on our worldview.

Bohr insisted on one aspect: two particles, once they became entangled as a result of a microscopic event, remain entangled even if they move away at sidereal distances. Looking at A, we influence the quantum state that includes A and B. The spin of B, therefore, is determined by the size of that of A, wherever the two particles are located. This particular aspect of the EPR paradox would have been better understood thirty years later, thanks to the enlightening work of John Bell to which we will return. For now, know that the key word is "non-locality", another version of Einstein's nosy definition: "spectral action at a distance".

In classical physics, the state (A up, B down) is totally separate and distinct from (A down, B up). As we have seen before, everything is determined by the choice of the friend who packages the packages, and in principle the evolution of the system is known by anyone who examines the initial data. The two options are independent and opening the package only reveals which one was chosen. From a quantum point of view, however, the wave function that describes both A and B "entanglement" the options; when we measure A or B, the whole function changes simultaneously in every place in space, without anyone emitting observable signals that travel at speeds higher than the speed of light. That's it. Nature works in this way.

This authoritarian insistence may perhaps silence the doubts of a freshman physicist, but is it really enough to save our troubled philosophical souls? Surely Bohr's "refutation" did not satisfy Einstein and colleagues at all. The two contenders seemed to speak different languages. Einstein believed in classical reality, in the existence of physical entities with well defined properties, like electrons and protons. For Bohr, who had abandoned the belief in an independent reality, the "demonstration" of incompleteness by the rival was meaningless, because it was wrong just the way the other one conceived a "reasonable" theory. Einstein asked one day one of our colleagues: "But do you really believe that the Moon exists there only when you look at it? If in this question we replace our satellite with an electron, the answer is not immediate. The best way out is to bring up the quantum states and the probabilities. If we ask ourselves what is the spin of a certain electron emitted by a tungsten wire in an incandescent lamp, we know that it will be up or down with equal probability of 50%; if nobody measures it, it does not make sense to say that the spin is oriented in a certain direction. About Einstein's question, better to gloss. Satellites are much larger than particles.

But what kind of world do we live in?

We have dedicated this chapter to one of the most enigmatic aspects of quantum physics, the exploration of the micro-world. It would be already traumatic if it was a new planet subject to new and different laws of nature, because this would undermine the very foundations of science and technology, the control of which makes us rich and powerful (some of us, let's say). But what is even more disconcerting is that the strange laws of the micro-world give way to the old and banal Newtonian physics when the dimensional scale grows to the level of tennis balls or planets.

All the forces known to us (gravitation, electromagnetism, strong and weak interaction) are of local type: they decrease with distance and propagate at speeds strictly not higher than the speed of light. But one day a certain Mr. Bell came out, who forced us to consider a new type of interaction, non-local, which propagates instantly and does not weaken with increasing distance. And he also proved that it exists, thanks to the experimental method.

Does this force us to accept these inconceivable non-local distance actions? We are in a nice philosophical quagmire. As we understand how different the world is from our daily experience, we experience a slow but inevitable change of perspective. The last half century has been for quantum physics the accelerated version of Newton and Maxwell's long series of successes in classical physics. Certainly we have reached a deeper understanding of phenomena, since quantum mechanics is at the basis of all sciences (including classical physics, which is an approximation) and can fully describe the behavior of atoms, nuclei and subnuclear particles (quarks and leptons), as well as molecules, solid state, the first moments of life in our universe (through quantum cosmology), the great chains at the base of life, the frenetic developments of biotechnology, perhaps even the way human consciousness operates. It has given us so much, yet the philosophical and conceptual problems it brings with it continue to torment us, leaving us with a feeling of unease mixed with great hopes.

contemplated the strange lands of Bell.

Chapter 8

Quantum Physics in Present Times

In the previous chapters we have relived the stories of the brilliant scientists of the 20th century who built quantum physics among a thousand difficulties and battles. The journey has led us to follow the birth of fundamental ideas that seem revolutionary and anti-intuitive to those who know well classical physics, born with Galileo and Newton and refined over three centuries. In front of the many people there were many problems about the very nature of the theory, for example about the validity and limits of the Copenhagen interpretation (which still today some people challenge and try to deny). The majority of researchers, however, realized that they had in their hands a new and powerful tool to study the atomic and subatomic world and did not have too many scruples to use it, even if it did not match their ideas in the philosophical field. Thus new areas of research in physics were created, still active today.

Some of these disciplines have completely changed our way of life and greatly increased our potential to understand and study the universe. The next time one of you or a member of your family enters an MRI apparatus (hopefully never), consider this fact: while the machine buzzes, spins, advances the couch and makes sounds like a supernatural orchestra, while on a monitor in the control room a detailed image of your organs is formed, you are experimenting in an essential way the effects of applied quantum physics, a world of superconductors and semiconductors, spin, quantum electrodynamics, new materials and so on. Inside a resonance machine you are, literally, inside an EPR-type experiment. And if the diagnostic apparatus is instead a PET, a positron emission tomography, know that they are bombarding you with antimatter!

Overcoming the Copenhagen stalemate, quantum mechanics techniques have been used to address many practical and specific problems in areas that were previously considered intractable. Physicists have begun to study the mechanisms that govern the behavior of materials, such as how phase changes from solid to liquid to gas, or how matter responds to magnetization, heating and cooling, or why some materials are better conductors of electricity than others. All this falls largely within the so-called "condensed matter physics". To answer the above questions it would be sufficient to apply Schrödinger's equation, but over time more refined mathematical techniques have been developed, thanks to which we have been able to design new and sophisticated toys, such as transistors and lasers, on which the entire technology of the digital world we live in today is based.

Most of the applications of colossal economic value derive from quantum electronics or condensed matter physics and are "non-relativistic", i.e. they do not depend on Einstein's theory of narrow relativity because they involve phenomena that occur at lower speeds than light. The same Schrödinger's equation is non-relativistic, and in fact it provides an excellent approximation of the behavior of electrons and atoms at not high speed, which is true both for the external electrons of the atoms, chemically active and involved in the bonds, and for the electrons moving inside solids.[1]

But there are open questions that involve phenomena that occur at speeds close to the speed of light, for example: what holds the nucleus together? what are the really fundamental building blocks of matter, the real elementary particles? how do you reconcile narrow relativity with quantum theory? We must then enter a faster world, different from the world of physics of materials. To understand what happens in the nucleus, place where mass can be converted into energy as in the case of radioactive decay (fission or fusion), we must consider quantum phenomena that occur at speeds close to that of light and go into the rough terrain of the theory of narrow relativity. Once we understand how things work, we can take the next step towards the most complicated and deep general relativity, which deals with the force of gravity. And finally face the problem of problems, that remained open until the end of the Second World

War: how to describe in a complete way the interactions between a relativistic electron (i.e. fast) and light.

The marriage between quantum physics and narrow relativity

Einstein's theory is the correct version of the concept of relative motion, generalized at speeds even close to those of light. Basically, it postulates general principles related to the symmetry of physical laws2 and has profound implications on particle dynamics. Einstein discovered the fundamental relationship between energy and momentum, which differs radically from the Newtonian one. This conceptual innovation is at the basis of the changes that must be made to quantum mechanics to reformulate it in a relativistic sense.3

A question then arises spontaneously: what arises from the marriage of these two theories? Something extraordinary, as we will see shortly.

$E = mc^2$

The equation $E=mc^2$ is very famous. You can find it everywhere: on t-shirts, in the graphic design of the television series Ai confini della realtà, in certain trademarks and in endless cartoons of the "New Yorker". It has become a sort of universal emblem of all that is scientific and "intelligent" in contemporary culture.

Rarely, however, do some television commentators bother to explain its true meaning, except to summarize it in the expression "mass is equivalent to energy". Nothing could be more wrong: mass and energy are actually completely different. Photons, just to give an example, do not have mass but can easily have energy.

The real meaning of E=mc2 is actually very specific. Translated in words, the equation tells us that "a resting particle of mass m has an energy E whose value is given by the relation E=mc2". A massive particle, in principle, can spontaneously transform itself into other lighter particles in a process (decay) that involves the release of energy.4 Nuclear fission, in which a heavy atomic nucleus breaks giving rise to lighter nuclei, as in the case of U235 (uranium-235), therefore produces a lot of energy. Similarly, light nuclei such as deuterium can combine in the nuclear fusion process to form helium, releasing also in this case large amounts of energy. This happens because the mass of the sum of the two starting nuclei is greater than that of the helium nucleus. This process of mass-energy conversion was simply incomprehensible before Einsteinian relativity, yet it is the engine that makes the Sun work and is the reason why life, beauty, poetry exist on Earth.

When a body is in motion, the famous formula E = mc2 must be modified, as Einstein himself knew well.5 Actually, the static formula (body with zero momentum) is deduced from the dynamic formula (body with non-zero momentum) and its form is not the one everyone knows, but this one:

$$E2 = m2c4$$

It would seem to be a matter of goat's wool, but in reality there is a big difference, as we will illustrate now. To get the energy of a particle we have to take the square root of the two members of this equation and we find the familiar E=mc2. But not only!

It is a simple arithmetic fact: numbers have two square roots, one positive and one negative. The square root of 4, for example, is both √4=2 and √4=-2, because we know that 2×2=4 but also (-2)×(-2)=4 (we know that two negative numbers multiplied together result in a positive number). So also the equation written above, solved with respect to E, gives us two solutions: E=mc2 and E=-mc2.

Here is a nice enigma: how can we be sure that the energy derived from Einstein's formula is always positive? Which of the two roots should we take? And how does nature know?

At first the problem didn't seem too serious, but it was stolen as a useless and silly sophistication. Those who knew it were in no doubt, the energy was always either nothing or positive, and a particle with negative energy was an absurdity that should not even be contemplated, on pain of ridicule. Everybody was too busy playing with Schrödinger's equation, that in its original form applies only to slow particles, like the external electrons of atoms, molecules and solid bodies in general. In its non-relativistic version, the problem does not arise because the kinetic energy of moving particles is always given by a positive number. And common sense leads us to think that the total energy of a mass particle at rest is mc2, that is it is positive too. For these reasons, physicists of the time did not even consider the possibility to consider the negative square root and labeled that solution as "spurious", that is "not applicable to any physical body".

But let's suppose for a moment that instead there are particles with negative energy, corresponding to the solution with the minus sign in front and that is to say with an energy at rest equal to -mc2. If they were to move, the energy would increase in modulus and thus become even less as the momentum increases.6 In possible collisions with other particles would continue to lose energy, as well as due to the emission of photons, and therefore their speed would increase more and more, getting closer to that of light. The process would never stop and the particles in question would have an energy that would tend to become infinite, or rather infinitely negative. After a while the universe would fill up with these oddities, particles that radiate energy constantly sinking deeper and deeper into the abyss of negative infinity.7

The square root century

It is truly remarkable that one of the fundamental engines of physics in the 20th century is the problem of "getting the right square root". With hindsight, the construction of quantum theory can be seen as the clarification of the idea of "square root of a probability", the result of which is Schrödinger's wave function (whose square, we remember, provides the probability of finding a body in a certain place and at a certain time).

The simple root extraction can give rise to real oddities. For example we obtain objects called imaginary and complex numbers, which have a fundamental role in quantum mechanics. We have already met a notorious example: $i = \sqrt{-1}$, the root of minus one. Quantum physics must necessarily involve i and his brothers because of its mathematical nature and there is no way to avoid them. We have also seen that the theory foresees oddities such as entanglement and mixed states, which are "exceptional cases", consequences due to the fact that everything is based on the square root of probability. If we add and subtract these roots before elevating everything to the square we can obtain term cancellations and therefore a phenomenon like interference, as we have seen from Young's experiment onwards. These oddities challenge our common sense as much as, and perhaps more so, the square root less one would have seemed absurd to the cultures that preceded us, like the ancient Greeks. At the beginning they did not even accept irrational numbers, so much so that according to a legend Pythagoras condemned his disciple who had demonstrated the irrationality of $\sqrt{2}$, that is the fact that this number cannot be written in the form of a fraction, a ratio between two integers. At the time of Euclid things had changed and the irrationality accepted, but as far as we know the idea of imaginary numbers was never contemplated (for more details see note 15 of chapter 5).

Another sensational result obtained by the physics of the last century can be considered a consequence of this simple mathematical structure, namely the concept of spin and spinor. A spinor is in practice the square root of a vector (see Appendix for more details). A vector, which is perhaps more familiar to you, is like an arrow in space, with defined length and direction, representing quantities such as, for example, the speed of a particle. Taking the square root of an object with spatial direction seems a strange idea and in fact has strange consequences. When you rotate a spinor $360°$ it does not return equal to itself, but becomes less itself. The calculations tell us then that if we exchange the position of two identical electrons with spin 1/2, the wave function of the state that includes the position of both must change sign: $\Psi(x,y) = -\Psi(y,x)$. The Pauli exclusion principle comes from this very fact: two identical particles with spin 1/2 cannot have the same state, because otherwise the wave function would be identically null. We have already seen that the principle applied to electrons leads us to exclude the presence of more than two particles on an orbital, one of which has spin up and the other spin down. Hence the existence of a repulsive "exchange force" between two particles with spin 1/2 that do not want at all costs to be in the same quantum state, which includes being in the same place at the same time. Pauli's exclusion principle governs the structure of the periodic table of elements and it is a very visible and fundamental consequence of the incredible fact that electrons are represented as square roots of vectors, that is spinors.

Einstein's formula that binds mass and energy gives us another situation in which the physics of the twentieth century had to compare with square roots. As we said, at the beginning everybody ignored the problem neglecting the negative solutions in the study of particles as photons or mesons. A meson is a particle with zero spin, while the photon has spin equal to 1, and their energy is always positive. In the case of electrons, that have spin 1/2, it was necessary to find a theory that integrate quantum mechanics and narrow relativity; and in this field we were face to face with negative energy states, that here give us the opportunity to know one of the most important figures of physics of the twentieth century.

Paul Dirac

Paul Dirac was one of the founding fathers of quantum physics, author, among other things, of the sacred book of this discipline: The Principles of Quantum Mechanics.[8] It is a reference text that deals coherently with the theory according to the Bohr-Heisenberg school of thought and combines Schrödinger's wave function with Heisenberg's matrix mechanics. It is a recommended reading for those who want to go deeper into the subject, even if it requires knowledge at the level of the first years of university.

Dirac's original contributions to 20th century physics are of paramount importance. Noteworthy, for example, is his theoretical proposal on the existence of magnetic monopoles, the magnetic field equivalent of electric charges, point sources of the field itself. In Maxwell's classical theory this possibility is not contemplated, because magnetic fields are considered generated only by charges in motion. Dirac discovered that monopoles and electric charges are concepts not independent but related through quantum mechanics. His theoretical speculations combined the new physics with a branch of mathematics called topology, which in those years was gaining importance. Dirac's theory of magnetic monopolies had a strong resonance also from a strictly mathematical point of view and in many ways anticipated the conceptual framework that was later developed by string theory. But his fundamental discovery, one of the deepest in 20th century physics, was the relativistic theory of the electron.

In 1926 the young Dirac was looking for a new equation to describe the spin 1/2 particles, that could overcome Schrödinger's one and take into account the restricted relativity. To do this he needed the spinors (the square roots of the vectors, remember) and he had to assume that the electron had mass. But in order to get back to relativity, he discovered that he had to double the spinors compared to the non-relativistic situation, therefore he had to assign two spinors to each electron.

In broad lines a spinor is composed by a couple of complex numbers that represent respectively the root of the probability to have spin up or spin down. In order to make this tool fall in the narrow relativity range, Dirac found a new relation where four complex numbers were needed. This is known today, you may have imagined, as Dirac's equation.

Dirac's equation takes the square roots really seriously, in the maximum generality. The two starting spinors represent two electrons, one spin up and one spin down, that without taking into account relativity have positive energy, so from $E^2=m^2c^4$ we use only the solution $E=+mc^2$. But if instead we take relativity into account we need two other spinors, to which we associate the negative solution of Einstein's equation $E=-mc^2$. So they have negative energy. Dirac himself could not do anything about it, because this choice was obligatory if we wanted to take into account the symmetry requirements of restricted relativity, essentially referred to the correct treatment of the motions. It was frustrating.

The problem of negative energy is inextricably present in the very heart of narrow relativity and therefore cannot be ignored. Dirac noticed how it became more and more thorny as he was progressing in his quantum electron theory. The minus sign cannot be ignored saying it is a not admissible solution, because the theory that results from the combination of quantum and relativity allows particles to have both positive and negative energy. We could settle the matter stating that an electron of negative energy is only one of the many "allowed quantum states", but this would lead to the disaster: the hydrogen atom, and all the ordinary matter, would not be stable. An electron of positive energy mc^2 could emit photons with energy equal to

2mc2, become a particle of negative energy -mc2 and begin the descent into the abyss of the less infinite (as the momentum increases the negative energy modulus would increase quickly).9 The universe would not be stable if negative energy states really exist. These new solutions with the minus sign in front were a real thorn in the side of the nascent theory.

However, Dirac had a brilliant idea to solve the problem. As we have seen, Pauli's exclusion principle states that two electrons cannot have exactly the same quantum state at the same time: if one is already in a certain state, like for example in an atomic orbital, nobody else can occupy that place (of course we must take into account also the spin, therefore in a state with certain characteristics of location and motion two electrons can live together, one with spin up and the other with spin down). Dirac had the idea to extend it to the vacuum: also the vacuum is actually full of electrons, that occupy all the negative energy states. These problematic states in the universe are therefore all occupied by two electrons, one spin up and the other spin down. In this configuration, the positive energy electrons of the atoms could not emit photons and find themselves in a negative energy state, because they would not find any free and thanks to Pauli's principle the action would be forbidden to them. With this hypothesis the vacuum would become analogous to a giant inert atom, like the one of a noble gas, with all the orbital full, that is with all the negative energy states occupied, for any amount of motion.

The supersymmetry

The calculation of vacuum energy continues to get worse if we stick muons, neutrinos, tau, quarks, gluons, W and Z bosons, the brand-new Higgs boson - i.e. all the particles that inhabit Mother Nature's zoo. Each of them provides a piece of total energy, positive for fermions and negative for bosons, and the result is always uncontrollable, i.e. infinite. Here the problem is not finding a better way to do the calculations but a new general principle that tells us how to derive the energy density of the vacuum in the universe. And so far we do not have it.

However, there is a very interesting symmetry that, if implemented in a "adjusted" quantum theory, allows us to calculate the cosmological constant and to obtain a mathematically comforting result: zero. This symmetry is given by a particular connection between fermions and bosons. To see it at work we must introduce an extra imaginary dimension on the scene, something that may have come out of the imagination of Lewis Carroll. And this additional dimension behaves like a fermion: with a principle "à la Pauli" prohibits more than one step in it.

Wherever we enter the new dimension, we must immediately stop (it is like putting an electron in a quantum state: then we cannot add more). But when a boson puts its foot in it, it is transformed in a fermion. And vice versa. If this strange dimension really exists and if we enter in it, like Alice in Through the mirror, we would see the electron transform in a boson called selectron and the photon become a fermion called photon.

The new dimension represents a new type of physical symmetry, mathematically consistent, called super symmetry, that associates to each fermion a boson and vice versa.15 In practice, the particle zoo doubles. The relationship between supersymmetric partners is similar to that between matter and antimatter. As you may have guessed, the presence of these new particles has a pleasant effect on the calculation of vacuum energy: the positive values of the bosons cancel the negative values of the fermions coming from the sea of Dirac, and the result is a nice zero. The cosmological constant is identically null.

So supersymmetry solves the very concrete problem of vacuum energy? Maybe so, but it is not very clear how. There are two obstacles. First, no bosonic supersymmetric partner of the electron has been observed yet.16 Second, there is evidence that the cosmological constant is actually small but different from zero. But the super symmetry, like all symmetries (think about a perfectly spherical glass bowl), can be "broken" (just give a hammer to this sphere). Physicists have a deep love for symmetry, which is always an ingredient of our most appreciated and used theories. So many colleagues hope that super symmetry really exists in nature and that there is also a mechanism (similar to the hammering on the sphere) that can break it. If so, we would observe its consequences only at very high energies, like the ones we hope to obtain in colossal accelerators like LHC.

According to the theory, the partners of the photon electron, that is selectortron and photon, are very heavy and we will be able to see their traces only when we will reach an energy equal to a threshold value called ΛSUSY (SUSY is the abbreviation of super symmetry).

Unfortunately, the breaking of super symmetry brings the problem of infinite energy back on the scene. The vacuum energy density is given by the formula seen above, i.e. $\Lambda 4SUSY/\hbar 3c3$. If we suppose that this quantity is of the order of magnitude of the maximum energies that can be obtained by the big accelerators, from one thousand to ten thousand billion electron volts, then we obtain a cosmological constant 1056 times bigger than the one observed. It is a remarkable improvement compared to 10120 of before, but it is still a beautiful problem. Supersymmetry, therefore, if directly applied to calculations, does not solve the crisis of vacuum energy. We must try other ways.

The holographic principle

Do we make any mistakes when counting the fish caught in the sea of Dirac? Perhaps we are raising too many? After all we are adding up really small creatures, negative energy electrons with very short wavelengths. The scale is really microscopic, even if we set a high value for the threshold energy Λ. Maybe these states are not really to be considered?

In the last ten years or so a new and radical hypothesis has emerged, according to which until now we have always overestimated the number of fish in the Dirac Sea, because these are not three-dimensional objects in a three-dimensional ocean, but are part of a hologram. A hologram is a projection of a certain space on another smaller one, as it happens when we project a three-dimensional scene on a two-dimensional screen. According to this hypothesis, everything that happens in three dimensions can be described completely based on what happens on the screen, with one dimension less. The Sea of Dirac, it follows, is not full of fish in the way we think, because these are actually two-dimensional objects. In short, the negative energy states are a mere illusion and the total energy of the vacuum is much less, so much so that it is potentially compatible with the observed value of the cosmological constant. We say "potentially" because holographic theory is still a construction site and still has many open points.

This new idea comes from the field of string theory, in areas where connections between holographic spaces can be established (the most defined and original is given by the so-called Maldacena conjecture, or AdS/CFT).17 We will come back to this hypothesis, unrealistic or deep, in the next chapter; however, the general sense seems to be that of a dreamer's logic.

The physics of condensed matter

Quantum theory allows interesting and very useful applications to materials science. For a start, it has allowed us for the first time in history to really understand what are the states of matter, how phase transitions work and electrical and magnetic properties. As in the case of the periodic table of the elements, quantum mechanics has largely repaid the investments made, both from a theoretical and practical point of view, thanks to the birth of new technologies. It has allowed the start of the so-called "quantum electronics" and has revolutionized the life of all of us in ways that were once inconceivable. Let's take a look at one of the main subsectors of this discipline, which deals with the flow of electric current in materials.

The conduction band

When atoms join together to form a solid, they find themselves crushed a short distance from each other. The wave functions of the external electrons, located in the higher level occupied orbitals, start in a way to merge (while the ones in the internal orbitals remain substantially undisturbed). The external electrons can jump from one atom to another, so much that the external orbital lose their identity and are no longer located around a specific atom: they merge then in a collection of extended electronic states that is called valence band.

Let's take for simplicity a crystalline material. The crystals present themselves in many forms, whose properties have been catalogued in detail.19 The electrons that begin to wander among the crystals have wave functions with very low frequency in the valence band. Within this band, the electrons are arranged following the Pauli exclusion principle: at most two per level, one with spin up and the other with spin down. The states with very low frequency are very similar to those of an electron free to move in space, without interference with the crystalline lattice. These states have the lowest energy level and fill up first. Always following the exclusion principle, electrons continue to fill the following states until their quantum wavelengths shorten, becoming comparable to the distance between atoms.

The electrons, however, are subject to deviations from the electromagnetic field generated by the crystal lattice atoms, which behaves like a giant interferometer of Young with a lot of cracks: one for each atom (the analogous of the cracks are the lattice charges). The motion of these particles, therefore, involves a colossal quantum interference.[20] This intervenes just when the wavelength of the electron is of the same order of magnitude compared to the distance between the atoms: the states in this condition are subject to destructive interference and therefore they cancel themselves.

The interference causes the formation of bands in the structure of energy levels of electrons in solids. Between the minimum energy band, the valence band, and the next one, called conduction band, there is a gap called energy gap. From this band structure directly depends the electrical behavior of the material. We can give three different cases.

1. Insulators. If the valence band is full and the energy gap before the conduction band is consistent, the material is called insulation. Insulators, such as glass or plastic, do not conduct

electricity. A typical representative of this category is an element that has almost all the orbitals full, such as halogens and noble gases. In these circumstances, the electric current does not flow because there is no space for electrons in the valence band to "wander". So in order to move freely an electron should jump to the conduction band, but if the energy gap is too much more this requires too much energy.21

2. Conductors. If the valence band is not completely full, then the electrons can move easily and enter new states of motion, which makes the material a good conductor of electric currents. The typical member of this category has a lot of electrons available to escape in the external orbital, which are not complete; they are therefore elements that tend to give electrons in chemical bonds, like alkali metals and some heavy metals. Among other things, it is the light diffusion by these free electrons to cause the typical shiny appearance of metals. As the conduction band fills up, the material becomes a less and less efficient conductor, until it reaches the insulating condition.

3. Semiconductors. If the valence band is almost complete and the conduction band has few electrons, the material should not be able to conduct much current. But if the energy gap is not too high, say less than about 3 eV, it is possible to force electrons without too much effort to jump in the conduction band. In this case we are in presence of a semiconductor. The ability of these materials to conduct electricity only in appropriate circumstances makes them really useful, because we can

manipulate them in various ways and have real "electronic switches" available.

Typical semiconductors are crystalline solids such as silicon (the main component of sand). Their conductivity can be modified drastically with the addition of other elements as "impurities", with a technique called doping. Semiconductors with few electrons in the conduction band are called "n" type and usually are doped with the addition of atoms that give up electrons, which go to "repopulate" the conduction band. Semiconductors with almost full conduction band are called p type and usually are doped with the addition of atoms that accept electrons coming from the valence band.

A p type material has "holes" in the valence band that look very similar to the ones we met when we talked about the Dirac sea. In that case we saw that these gaps were acting as positive particles; here something similar happens: the holes, called gaps, take the role of positive charges and facilitate the passage of electric current. A p-type semiconductor, therefore, is a kind of miniature Dirac sea, created in a laboratory. In this case however the gaps involve a lot of electrons and behave as if they were heavier particles, therefore they are less efficient as current carriers than the single electrons.

Diodes and transistors

The simplest example of a mechanism that we can build thanks to semiconductors is the diode. A diode acts as a conductor in one direction and an insulator in the other. In order to build one, in general, a p-type material is coupled with an n-type material to form a p-n junction. It does not take much effort to stimulate the electrons of type n element to go beyond the junction and end up in the valence band of type p element. This process, which resembles the particle-anti-particle annihilation in the Dirac sea, makes the current flow only one direction.

If we try to reverse the phenomenon we realize that it is difficult: removing electrons from the conduction band we do not find one able to replace them coming from p type material. In a diode, if we do not exaggerate with the voltage (it is easy to burn a semiconductor with a too intense current), we can easily make electricity flow in one direction only, and that is why this device has important applications in electric circuits.

In 1947 John Bardeen and William Brattain, who worked at Bell Laboratories in a group led by William Shockley, built the first "point contact" transistor. It was a generalization of the diode, made by joining three semiconductors. A transistor allows us to control the current flow as the voltage varies between the three layers (called respectively collector, base and emitter) and basically serves as a switch and amplifier. It is perhaps the most important mechanism invented by man and earned Bardeen, Brattain and Shockley the Nobel Prize in 1956.22

Profitable Applications

In the end what comes in our pockets? Schrödinger's fundamental equation, which provides us with a way to calculate the wave function, was born as the birth of pure reason: nobody imagined then that it would be the basis for the operation of expensive machinery or would feed the economy of a nation. But if applied to metals, insulators and (the most profitable) semiconductors, this equation has allowed us to invent particular switches and control mechanisms that are indispensable in equipment such as computers, particle accelerators, robots that build cars, video games and airplanes able to land in any weather.

Another favorite son of the quantum revolution is the ubiquitous laser, which we find in supermarket checkouts, eye surgery, precision metallurgy, navigation systems and the instruments we use to probe the structure of atoms and molecules. The laser is like a special flashlight that emits photons all of the same wavelength.

We could go on and on about the technological miracles that owe their existence to the intuitions of Schrödinger, Heisenberg, Pauli and many others. Let's see some of them. The first one that comes to our mind is the tunnel effect microscope, able to reach magnifications thousands of times higher than the most powerful electron microscope (which is in turn son of the new physics, because it is based on some wave characteristics of electrons).

The tunnel effect is the quintessence of quantum theory. Imagine a metal bowl placed on a table, inside of which there is a steel ball free to roll without friction. According to classical physics the sphere is trapped in the bowl for eternity, condemned to go up and down along the walls always reaching the same height. Newtonian system to the highest degree. For the quantum version of this configuration, we take an electron confined in a metal cage, in whose walls circulates current with a voltage that the particle does not have enough energy to counteract. So the electron approaches to a wall, it is rejected to the opposite wall, it is still rejected, and so on to infinity, right? No! Sooner or later, in the strange quantum world, the particle will be outside.

Do you understand how disturbing this is? In the classic language we would say that the electron has magically dug a tunnel and escaped from the cage, as if the metal sphere, novella Houdini, had escaped from the prison of the bowl. Applying to the problem Schrödinger's equation we make you enter the probability: at every encounter between electron and wall there is a small possibility that the particle goes beyond the barrier. Where does the necessary energy go? Not relevant question, because the equation only tells us what is the probability with which the electron is inside or outside. For a Newtonian mind this doesn't make sense, but the tunnel effect has very tangible effects. In the forties of the last century had become a fact that physicists used to explain nuclear phenomena previously incomprehensible. Some parts of the atomic nucleus manage, due to the tunnel effect, to cross the barrier that keeps them bound and in doing so they break the original nucleus to form smaller ones. This is fission, phenomenon at the base of nuclear reactors.

Another apparatus that puts this strange effect into practice is the Josephson junction, a sort of electronic switch so named in honor of its brilliant and bizarre inventor, Brian Josephson. This device operates at temperatures close to absolute zero, where quantum superconductivity adds an exotic character to the phenomena. In practice it is a super-fast and super-cool digital electronic apparatus that takes advantage of the quantum tunnel effect. It seems to come straight out of the pages of a book by Kurt Vonnegut, but it really exists and is able to turn on and off thousands of billions of times per second. In our era of increasingly powerful computers, this speed is a very desirable feature. This is because the calculations are done on bits, that is on units that can be zero or one, thanks to various algorithms that represent all the numbers, add them up, multiply them, calculate derivatives and integrals, and so on. All this is done by changing the value of certain electrical circuits from zero (off) to one (on) several times, so you will understand that speeding up this operation is of paramount importance. The Josephson junction does it better than any other.

The tunnel effect applied to the microscopy has allowed us to "see" the single atoms, for example in the fantastic architecture of the double helix that forms the DNA, register of all the information that defines a living being. The tunnel effect microscope, invented in the eighties of the last century, does not use a light beam (as in the optical microscope) or an electron beam (as in the standard electronic one). Its operation is based on a very high precision probe that follows the contour of the object to be observed remaining at a distance of less than one millionth of a millimeter. This gap is small enough to allow the electric currents present on the surface of the object itself to overcome it thanks to the tunnel effect and to stimulate a very sensitive crystal present in the probe. Any variation in this distance, due to an atom "protruding", is recorded and converted into an image by a special software. It is the equivalent of the stylus of a turntable (does anyone remember?), which runs through hills and valleys in a groove and turns them into Mozart's magnificent music.

The tunnel effect microscope is also able to take atoms one by one and move them elsewhere, which means we can build a molecule according to our designs, putting the pieces together as if it were a model. It could be a new very resistant material or an antiviral drug. Gerd Benning and Heinrich Rohrer, who invented the tunnel effect microscope in an IBM laboratory in Switzerland, were awarded the Nobel Prize in 1986, and their idea gave birth to an industry with a billionaire turnover.

In the present and the near future there are two other potential revolutions: nanotechnology and quantum computing. Nanotechnologies (where "nano" is the prefix that is worth 10-9, which means "really very small") are the reduction of mechanics, with motors, sensors and so on, at atomic and molecular scale. We literally speak of Lilliputian factories, where a reduced size a million times corresponds to an equal increase in operating speed. Quantum production systems could use the most raw materials of all, atoms. Our polluting factories would be replaced by compact, efficient machines.

Quantum computation, to which we will return later, promises to offer "a system of information processing so powerful that by comparison traditional digital computation will look like a smash compared to a nuclear reactor".

Chapter 9
The third millennium

As we have seen on several occasions during the course of the book, quantum science, despite its bizarre idea of reality, works very well, on an almost miraculous level. Its successes are extraordinary, profound and of great weight. Thanks to quantum physics we have a true understanding of what happens at molecular, atomic, nuclear and subnuclear level: we know the forces and laws that govern the microworld. The intellectual depth of its founders, at the beginning of the 20th century, has allowed us to use a powerful theoretical tool that leads to surprising applications, those that are revolutionizing our way of life.

From his cylinder, the quantum wizard has brought out technologies of unimaginable scope, from lasers to tunnel effect microscopes. Yet some of the geniuses who helped create this science, wrote the reference texts and designed so many miraculous inventions are still in the throes of anguish. In their hearts, buried in a corner, there is still the suspicion that Einstein was not wrong and that quantum mechanics, in all its splendor, is not the final theory of physics. Come on, how is it possible that probability is really part of the basic principles that govern nature? There must be something that escapes our attention. Gravity, for example, which has been neglected by the new physics for a long time; the dream of arriving at a solid theory unifying Einstein's general relativity and quantum mechanics has led some daredevil to probe the depths of the fundamentals, where only the most abstract mathematics provides a weak light, and to conceive string theory. But is there perhaps something even deeper, a component missing in the logical foundations of quantum physics? Are we trying to complete a puzzle that lacks a piece?

Some people ardently hope to arrive soon to a super-theory that is reduced to quantum mechanics in certain areas, as happens with relativity that engulf Newtonian classical mechanics and returns value only in certain areas, that is when the bodies in play move slowly. This would mean that modern quantum physics is not the end of the line, because over there, hidden deep in the mind of Nature, there is a definitive theory, better, able to completely describe the universe. This theory would be able to face the frontiers of high energy physics, but also the intimate mechanisms of molecular biology and complexity theory. It could also lead us to discover completely new phenomena, so far escaped from the eye of science. After all, our species is characterized by curiosity and is unable to resist the temptation to probe this exciting and surprising microworld as a planet orbiting a distant star. And research is also a big business, if it is true that 60% of the American GDP depends on technologies that have to do with quantum physics. So there are good reasons to continue to explore the foundations of the building on which we build our understanding of the world.

"Quantum phenomena challenge our primitive understanding of reality; they force us to re-examine the very idea of existence," writes Euan Squires in the preface to his book The Mystery of the Quantum World. "These are important facts, because our beliefs about "what exists" certainly influence the way we conceive our place in the world. On the other hand, what we believe we are ultimately influences our existence and our actions".1 The late Heinz Pagels in his essay The Cosmic Code speaks of a situation analogous to that of a consumer who must choose a variant of "reality" from among the many on offer at a department store.

We have challenged the common conception of "reality" in the previous chapters when we dealt with Bell's theorem and its experimental consequences, i.e. we have taken into consideration the possibility of non-local effects: the instantaneous transfer of information between two places located at an arbitrary distance. According to the classic way of thinking, the measurement made in one point "influences" the observations in the other; but in reality the link between these two places is given by a property of the two particles (photons, electrons, neutrons or whatever) that were born together in an entangled state. At their arrival in the points where the two detectors are located, if the apparatus 1 registers the property A for the receiving particle, it is necessary that the apparatus 2 registers the property B, or vice versa. From the point of view of wave functions, the act of measuring by apparatus 2 makes the quantum state "collapse" simultaneously at each point of space. Einstein hated this, because he firmly believed in location and the prohibition of exceeding the speed of light. Thanks to various experiments we have excluded the possibility that the two detectors exchange signals; the existence of entanglement is instead a known and widely confirmed fact, therefore once again quantum physics is correct at a fundamental level. The problem is all in our reaction to this phenomenon, apparently paradoxical. A theoretical physicist wrote that we should find a form of "peaceful coexistence" with quantum mechanics.

The crux of the question is then: is the EPR paradox an illusion, perhaps conceived in a way that seems deliberately anti-intuitive? Even the great Feynman felt challenged by Bell's theorem and tried to arrive at a representation of quantum mechanics that would make it more digestible, thanks to his idea of the sum on the paths. As we have seen, starting from certain ideas of Paul Dirac invented a new way of thinking about events. In his framework, when a radioactive particle decays and gives life to a couple of other particles, one with spin up and the other with spin down, we must examine the two "pathways" that are determined. One, which we will call A, brings the particle with spin up to detector 1 and the one with spin down to detector 2;

the other, path B, does the opposite. A and B have quantistically two "probability amplitudes", which we can add up. When we make a measurement, we also find out which of the two paths the system has actually taken, if A or B; so if for example we find the particle with spin up at point 1, we know that it has passed through A. In all this, we are only able to calculate the probability of the various paths.

With this new concept of space-time disappears the possibility that the information propagates instantly even at great distances. The general picture is closer to the classic one: you will remember the example in which our friend sends to us on Earth and to a colleague on Rigel 3 a colored ball, which can be red or blue; if opening the package we see the blue ball, we know in that same instant that the other received the red ball. Yet nothing changes in the universe. Perhaps this model calms our philosophical fears about the EPR paradox, but it has to be said that even the sum on the paths has some really disconcerting aspects. The mathematics behind it works so well that the model excludes the presence of signals that travel faster than light. This is closely related to facts such as the existence of antimatter and quantum field theory (as we have seen in chapter 8). We see therefore that the universe is conceivable as an infinite set of possible paths that govern its evolution over time. It is as if there was a giant wavefront of probability that advances. From time to time we make a measurement, we select a path for a certain event of space-time, but after that the great wave gives a shake and continues its race towards the future.

Generations of physicists have felt the frustration of not knowing what quantum theory they were using really was. Even today the conflict between intuition, experiments and quantum reality can be profound.

Quantum Cryptography

The problem of secure transmission of information is not new. Since ancient times military espionage has often used secret codes, which counter-espionage has tried to break. In Elizabethan times, the deciphering of a coded message was the basis of Mary Stuart's death sentence. According to many historians, one of the fundamental junctions of the Second World War occurred when in 1942 the Allies defeated Enigma, the secret code of the Germans considered "invincible".8 The "big game" of the spies also consists in understanding if the code was violated and in preparing a countermove by spreading false information.

Today, as anyone who keeps himself informed knows, cryptography is no longer a matter for spies and the military. When you enter your credit card information on eBay or Amazon's website, you assume that the communication is protected. But the companies of hackers and information terrorists make us realize that the security of trade, from email to online banks, is hanging by a fragile thread. The American government takes the problem seriously, so much so that billions of dollars are spent on it.

The most immediate solution is to introduce a cryptographic "key" that can be used by both sender and receiver. The standard way to make confidential information secure is to hide it in a long list of random numbers. But we know that spies, hackers and strange guys dressed in black with a heart of stone and a good knowledge of the computer world are able to understand how to distinguish information from noise.

This is where quantum mechanics comes into play, which can offer cryptography the services of its special form of randomness, so strange and wonderful that it constitutes an insurmountable barrier - and as if that were not enough, it is able to immediately report any attempt to intrude! Since the history of cryptography is full of "impenetrable" codes that at some point are penetrated by a superior technology, you are justified if you take this assertion with the right amount of skepticism (the most famous case is the one already mentioned of Enigma, the machine that in the Second World War encrypted Nazi transmissions and was considered unbeatable: the Allies managed to decrypt it without the enemy noticing).

Let's see a little more in detail how encryption works. The minimum unit of information that can be transmitted is the bit, an abbreviation of binary digit, that is "binary number". A bit is simply zero or one. If for example we flip a coin and decide that

0 represents head and 1 is tail, the result of each flip is a bit and a series of flips can be written like this:

101100010111010010101010111.

This is as far as classical physics is concerned. In the quantum world there is an equivalent of the bit that has been baptized qubit (if you think it has something to do with the "cubit", a traditional unit of measurement that also appears in the Bible, you are off track). Also it is given by a variable that can assume two alternative values, in this case the electron spin, equal to up or down, that take the place of 0 and 1 of the classic bit. So far nothing new.

But a qubit is a quantum state, so it can exist also in "mixed" as well as "pure" form. A pure state is not influenced by observation. If we measure the spin of an electron along the z axis, it will necessarily be up or down, depending on its direction. If the electron is taken at random, each of these values can occur with a certain probability. If the particle has been emitted in such a way that it has to assume a certain spin, the measurement will only record it without changing its state.

In principle we can therefore transmit information in the form of binary code using a collection of electrons (or photons) with predetermined spin up or down on the z axis; since they have all been "pure", a detector oriented along the same axis will read them without disturbing them. But the z-axis must be defined, it is not an intrinsic feature of space. So here is a secret information that we can send to the recipient of the message: how the axis along which the spin is oriented.

If someone tries to intercept the signal with a detector not perfectly parallel to our z, with its measurement it disrupts the electronic states and thus obtains a meaningless set of data (without realizing it). Our receivers who read the message realize instead that something has interfered with the electrons and therefore that there has been an attempt of intrusion: we know that there is a spy listening and we can take countermeasures. Vice versa, if the message arrives without problems we can be sure that the transmission has been done in a safe way. The key point of the story is that the attempt to intrude causes changes in the qubit status, of which both the sender and the receiver are aware.

The transmission of quantum states can also be used to securely transmit a "key", i.e. a very large causally generated number, which is used to decode the information in certain encrypted communication systems. Thanks to qubits, we know if the key is secure or compromised and can therefore take countermeasures. Quantum cryptography has been tested so far on messages transmitted a few kilometers away. It will take some time before it can be used in practice, because this requires large investments in the field of lasers of the latest generation. But one day we will be able to make the hassle of having to contest a purchase charged on our credit card in some distant country where we have never been.

Quantum Computers

However, there is a threat to the security of quantum cryptography, and it is the quantum computer, the number one candidate to become the supercomputer of the 21st century. According to the empirical law enunciated by Gordon Moore, "the number of transistors on a chip doubles every twenty-four months".10 As some joker has calculated, if automobile technology had progressed at the same pace as computer technology over the last thirty years, we would now have sixty gram-heavy cars that cost forty dollars, with a trunk of one and a half cubic kilometers, consuming almost nothing and reaching speeds of up to one and a half million kilometers per hour.

In the computer field, we have gone from vacuum tubes to transistors and integrated circuits in less time than a human life span. Yet the physics on which these tools are based, including the best available today, is classical. Using quantum mechanics, in theory, we should build new and even more powerful machines. They have not yet appeared in IBM's design office or in the business plans of Silicon Valley's boldest startups (at least as far as we know), but quantum computers would make the fastest of the classical ones look like little more than an abacus in the hands of a mutilated person.

Quantum computation theory makes use of the already mentioned qubits and adapts to non classical physics the classical theory of information. The fundamental concepts of this new science were established by Richard Feynman and others in

176

the early eighties and had a decisive boost thanks to the work of David Deutsch in 1985. Today it is an expanding discipline. The keystone has been the design of "logic gates" (computer equivalents of switches) that take advantage of quantum interference and entanglement to create a potentially much faster way of making certain calculations.

Interference, made explicit by double slit experiments, is one of the strangest phenomena in the quantum world. We know that just two slits made in a screen change the behavior of a photon passing through it in a bizarre way. The explanation given by the new physics brings into question the probability amplitudes of the various paths that the particle can follow, which if summed up properly provide the probability that it will end up on a certain region of the detector. If instead of two slits there were a thousand, the basic principle would not change and to calculate the probability that the light arrives at this or that point we should take into account all the possible paths. The complexity of the situation increases further if we take two photons and not just one, each of which has thousands of choices, which brings the number of total states to the order of millions. With three photons the states become of the order of billions, and so on. The complexity grows exponentially to the increase of the inputs.

The final result is perhaps very simple and predictable, but to make all these accounts is very little practical, with a classic calculator. The great idea of Feynman was to propose an analogic calculator that exploits quantum physics: we use real photons and we really perform the experiment, leaving in this way that nature to complete in a fast and efficient way that monstrous calculation. The ideal quantum computer should be able to choose by itself the type of measurements and experiments that correspond to the required calculation, and at the end of the operations translate the physical result into the numerical result. All this implies the use of a slightly more complicated version of the double slit system.

The incredible computers of the future

To give you an idea of how powerful these calculation techniques are, let's take a simple example and compare a classical situation with the corresponding quantum one. Let's start from a "3-bit register", i.e. a device that at each instant is able to assume one of these eight possible configurations: 000, 001, 010, 011, 100, 101, 110, 111, corresponding to the numbers 0, 1, 2, 3, 4, 5, 6, 7. A classical computer records this information with three switches that can be open (value 0) or closed (value 1). It is easy to see by analogy that a 4-bit register can encode sixteen numbers, and so on.

However, if the register is not a mechanical or electronic system but an atom, it is able to exist in a mixed state, superimposing the fundamental one (which we make correspond to 0) and the excited one (equal to 1). In other words, it is a qubit. A 3 qubit register therefore expresses eight numbers at the same time, a 4 qubit register expresses sixteen and in general a N qubit register contains 2N.

In classical computers the bit is generally given by the electric charge of a small capacitor, which can be charged (1) or not charged (0). By adjusting the current flow we can change the bit value. In quantum computers, instead, to change a qubit we use a light beam to put the atom in excited or fundamental state. This implies that at each instant, at each step of the calculation, the qubit can assume the values 0 and 1 at the same time. We begin to realize great potential.

With a 10 registers qubit we are able to represent at each instant all the first 1024 numbers. With two of them, coupled in a matching way, we can make sure that we have a table of 1024 × 1024 multiplications. A traditional calculator, although very fast, should perform in sequence more than a million calculations to get all those data, while a quantum computer is able to explore all the possibilities simultaneously and get the right result in a single step, without effort.

This and other theoretical considerations have led to believe that, in some cases, a quantum computer would solve in a year a problem that the fastest of classic machines would not finish before a few billion years. Its power comes from the ability to operate simultaneously on all states and to perform many calculations in parallel on a single operating unit. But there is a but (suspense: here would fit Also Sprach Zarathustra by Richard Strauss). Before investing all your savings on a Cupertino startup, you should know that a number of experts are sceptical about quantum computer applications (although they all agree that theoretical discussions on the subject are valuable for understanding certain fundamental quantum phenomena).

It is true that some important problems can be solved very well, but we are still talking about very different machines, designed for very specific situations, which will hardly replace the current ones. The classical world is another kind of world, and that is why we do not bring the broken machine to the quantum mechanic. One of the major difficulties is that these devices are very sensitive to interference with the outside world: if a single cosmic ray makes a qubit change state, all the calculation is blessed. They are also analog machines, designed to simulate a particular calculation with a particular process, and therefore lack the universality typical of our computers, in which run programs of various nature that make us calculate everything we want. It is also very difficult to build them in practice. In order for quantum computers to become reality and it is worth investing time and money in them, we will have to solve complex reliability problems and find usable algorithms.

One of these potentially effective algorithms is the factoring of large numbers (in the sense of their breakdown into prime factors, such as 21=3×7). From the classical point of view, it is relatively easy to multiply the numbers among them but it is generally very difficult to do the reverse operation, i.e. to find the factors of a colossus like:

3 204 637 196 245 567 128 917 346 493 902 297 904 681 379

This problem has important applications in the field of cryptography and is a candidate to be the spearhead of quantum computing, because it is not solvable with classical calculators.

We also mention the bizarre theory of the English mathematician Roger Penrose that concerns our consciousness. A human being is able to perform certain types of calculations at lightning speed, like a calculator, but he does it with very different methods. When we play chess against a computer, for example, we assimilate a large amount of sensory data and quickly integrate it with experience, to counteract a machine that works in an algorithmic and systematic way. The computer always provides correct results, the human brain sometimes does not: we are efficient but inaccurate. We have sacrificed precision to increase speed.

According to Penrose the feeling of being conscious is the coherent sum of many possibilities, i.e. it is a quantum phenomenon. So according to him we are for all intents and purposes quantum computers. The wave functions that we use to produce computational results are perhaps distributed not only in the brain but throughout the body. In his essay Shadows of the Mind, Penrose hypothesizes that the wave functions of consciousness reside in the mysterious microtubules of neurons. Interesting, to say the least, but a true theory of consciousness is still missing.

Be that as it may, quantum computation could find its raison d'être by shedding light on the role of information in basic physics. Maybe we will be able both to build new and powerful machines, and to reach a new way of understanding the quantum world, maybe more in tune with our changing perceptions, less strange, ghostly, disturbing. If this will really happen, it will be one of the rare moments in the history of science when another discipline (in this case the theory of information, or perhaps consciousness) merges with physics to shed light on its basic structure.

Grand finale

Let us conclude our story by summarizing the many philosophical questions waiting to be answered: how can light be both a particle and a wave? are there many worlds or is there only one? is there a truly impenetrable secret code? what is reality at the fundamental level? are the laws of physics regulated by many dice throws? do these questions make sense? the answer is maybe "we have to get used to these oddities"? where and when will the next great scientific leap forward take place?

We started from the mortal blow inflicted by Galileo to Aristotelian physics. We have moved into the clockwork harmony of Newton's classical universe, with its deterministic laws. We could have stopped there, in a real and metaphorical sense, in that comforting reality (even if without cell phones). But we did not. We have penetrated into the mysteries of electricity and magnetism, forces that only in the nineteenth century were united and woven into the fabric of classical physics, thanks to Faraday and Maxwell. Our knowledge seemed then complete, so much so that at the end of the century there were those who predicted the end of physics. All the problems worth solving seemed to be solved: it was enough to add a few details, which would certainly come within the classical theories. End of the line, we go down; physicists can bundle up and go home.

But there was still some incomprehensible phenomenon here and there. The burning embers are red, while according to calculations they should be blue. And why is there no trace of ether? Why can't we go faster than a ray of light? Perhaps the last word was not yet said. Soon, the universe would be revolutionized by a new and extraordinary generation of scientists: Einstein, Bohr, Schrödinger, Heisenberg, Pauli, Dirac and others, all enthusiastic about the idea.

Of course, the old and dear Newtonian mechanics continue to work well in many cases, such as the motion of planets, rockets, balls and steam engines. Even in the XXVII century a ball thrown in the air will follow the elegant classical parable. But after 1900, or rather 1920, or better still 1930, who wants to know how the atomic and subatomic world really works is forced to change head and enter the realm of quantum physics and its intrinsic probabilistic nature. A realm that Einstein never wanted to accept completely.

We know that the journey has not been easy. The omnipresent double slit experiment alone can cause migraines. But it was only the beginning, because afterwards came the dizzying plateaus of

Schrödinger's wave function, Heisenberg's uncertainty and Copenhagen's interpretation, as well as various disturbing theories. We met cats alive and dead at the same time, beams of light behaving like waves and particles, physical systems linked to the observer, debates on the role of God as the supreme dice player... And when everything seemed to make some sense, here come other puzzles: Pauli's exclusion principle, the EPR paradox, Bell's theorem. It is not material for pleasant conversations at parties, even for New Age adepts who often formulate a wrong version of it. But we have made strength and we have not given up, even in front of some inevitable equation.

We have been adventurous and we have granted audience to ideas so bizarre that they could be Star Trek episode titles: "Many Worlds", "Copenhagen" (which is actually also a play), "The Strings and M-theory", "The Cosmic Landscape" and so on. We hope that you enjoyed the trip and that now, like us, you have an idea of how wonderful and deeply mysterious our world is.

In the new century looms the problem of human consciousness. Perhaps it can be explained thanks to quantum states. Although not a few people think so, it is not necessarily so - if two phenomena are unknown to us, they are not necessarily connected.

The human mind has a role in quantum mechanics, as you will remember, and that is when measurement comes into play. The observer (his mind) interferes with the system, which could imply a role of consciousness in the physical world. Does the mind-body duality have anything to do with quantum mechanics? Despite all that we have recently discovered about how the brain encodes and manipulates information to control our behavior, it remains a great mystery: how is it possible that these neurochemical actions lead to the "self", to the "inner life"? How is it possible to generate the feeling of being who we are?

There is no lack of critics of this correlation between quantum and mind, including the discoverer of DNA Francis Crick, who in his essay Science and the Soul writes: "The self, my joys and sorrows, my memories and ambitions, my sense of personal identity and free will, are nothing more than the result of the activity of a colossal number of neurons and neurotransmitters.

We hope that this is only the beginning of your journey and that you will continue to explore the wonders and apparent paradoxes of our quantum universe.

CPSIA information can be obtained
at www.ICGtesting.com
Printed in the USA
LVHW052215211220
674812LV00030B/1781

9 781838 310257